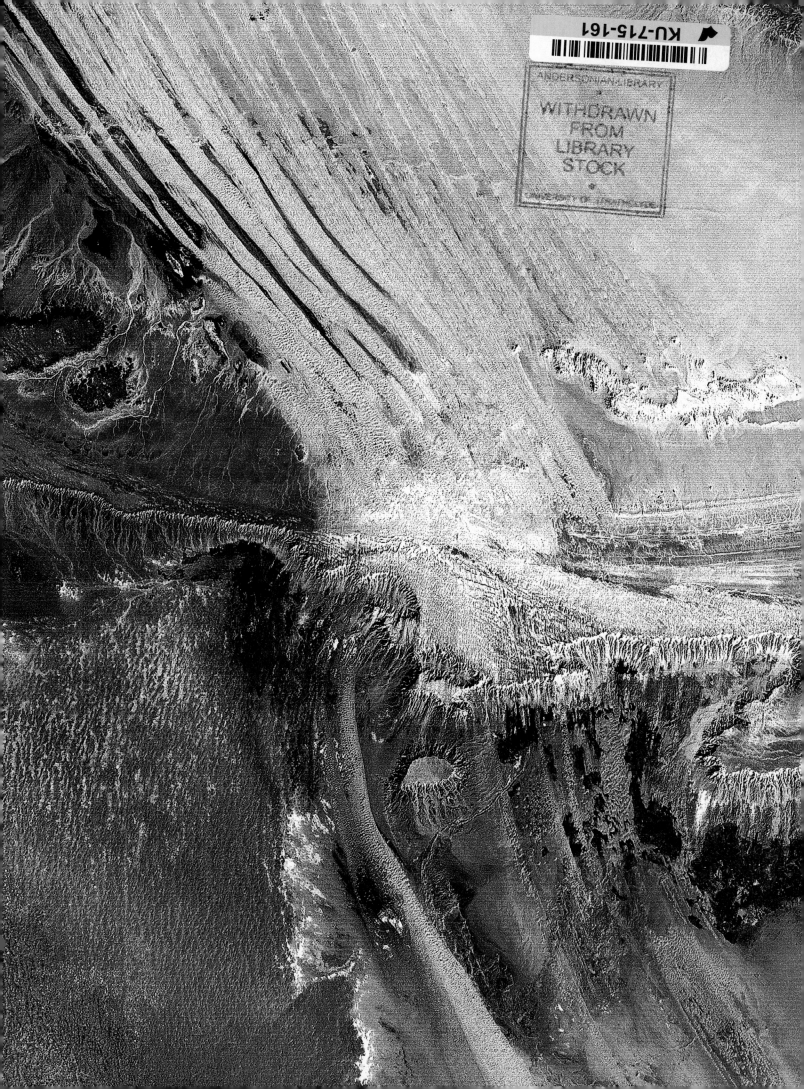

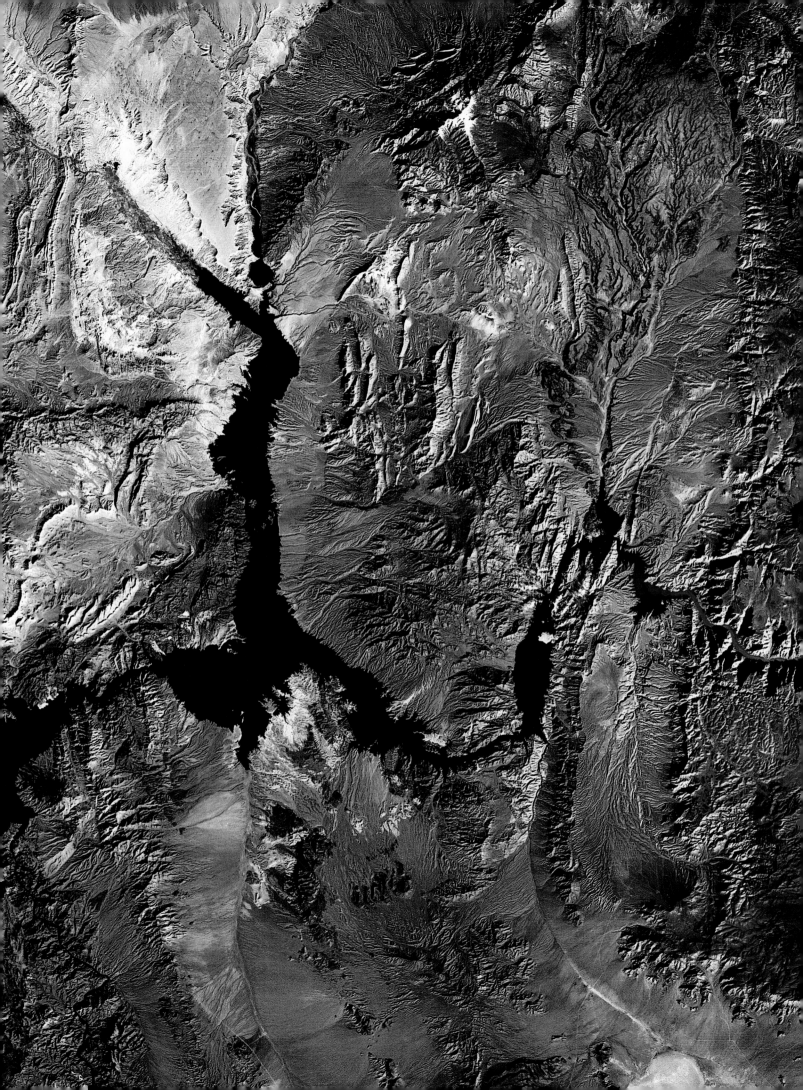

PRISCILLA STRAIN & FREDERICK ENGLE

Looking At Earth

A Smithsonian Columbus Quincentenary Project
Sponsored in part by the Eastman Kodak Company, Aerial Systems

Turner Publishing, Inc.

ATLANTA

copyright/staff credits

(previous pages)

Landsat images reveal the diversity of Earth's landforms, and show how remote sensing can capture scenes ranging from windswept deserts and volcanic islands to geologic structures and coastal urban areas.

The Western Desert of Egypt is a broad bedrock plain that is broken by depressions such as the Dakhla Oasis (see pages 2–3). Long dunes indicate that the prevailing winds near Dakhla are northerly.

A huge volcano that rises from the floor of the Pacific Basin forms the Big Island of Hawaii (see pages 4–5). Dark lava flows run down the slopes of Mauna Loa, and rich vegetation (deep red) grows on the northeastern side of the island (bottom).

Las Vegas (see pages 6–7) lies in a semiarid basin to the west of Lake Mead, which was formed by the damming of the Colorado River.

East of Providence, Rhode Island, Cape Cod (see pages 8–9) juts into the Atlantic, where its "forearm" is molded by ocean currents.

ISBN: 1-57036-373-0

Published by Turner Publishing, Inc.
A Subsidiary of Turner Broadcasting System, Inc.
1050 Techwood Drive, NW
Atlanta, Georgia 30318

Designed and produced on Macintosh computers using QuarkXPress, Aldus FreeHand, and Adobe Photoshop. Printed in Hong Kong.

First Edition

10 9 8 7 6 5 4 3 2 1

Editorial:

Katherine Buttler
EDITOR

Marian G. Lord
COPY EDITOR

Design:

Michael J. Walsh
DESIGN DIRECTOR

Karen E. Smith
BOOK DESIGN AND PRODUCTION

Andrew Johnston
COMPUTER CARTOGRAPHY

Gwenda L. Hyman
Barbara Griffin
PICTURE RESEARCH

Marty Moore
ASSISTANT PICTURE RESEARCH

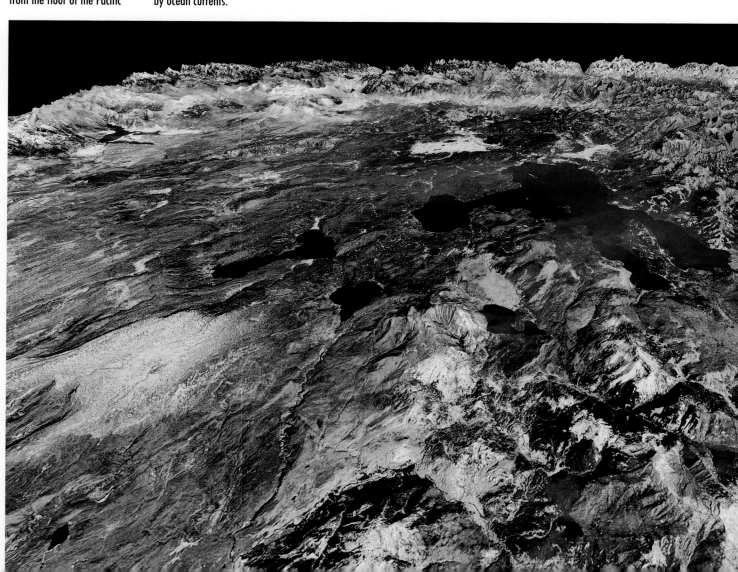

CONTENTS

(cover photo)

While devastating forest fires swept Yellowstone National Park in September 1988, a satellite passed overhead. A Landsat Thematic Mapper image was combined with topographic data to produce this computer-generated scene. Yellow spots indicate active fires. Large orange patches mark burned-over areas. Effects of smoke and haze have been reduced by computer image-processing techniques.

IMAGE COURTESY OF TASC, EOSAT CORP., AND CIRRUS TECHNOLOGY.

LOOKING AT EARTH FROM ORBIT IS AN AMAZING AND AWE-INSPIRING EXPERIENCE, ONE THAT SEEMS to change people in significant ways. Imagine soaring over the planet at an altitude of several hundred miles, traveling at a speed of nearly 18,000 miles an hour, yet feeling that you are floating in a great airship. Every moment presents you with a view that draws out myriad streams of thought, creating a wonderful weave of surprise and remembrance, curiosity and childlike delight. Colors you've never imagined, shapes that are strikingly like the maps you've seen so often before, places you've lived in, visited, or read about—they all scroll below at a seemingly majestic, yet, in actuality, startlingly fast pace.

The views before you cover a mind-boggling range of scales. You have a panoramic view of the planet, stretching across an arc of 180° and extending to a horizon nearly 1,000 miles away. Simultaneously, you can see immediately below you much smaller features: airports, towns, farm fields, the contrails of airplanes, and the wakes of ships.

At night, the most profuse array of stars you have ever seen fills your view toward the horizon. Every orbit gives you the treat of seeing constellations from another hemisphere—a rarity for most Earth-bound people. The thin, diffusely glowing band that marks the Earth's air-glow layer is plainly visible if the Moon is not too full. The air-glow curves away toward the pole, where it merges with the spectacular green arcs of the aurora. The arcs undulate in slow, grand waves, as if they were gigantic draperies stirred by a faint breeze. The more active arcs are higher and brighter, with rays that extend from beneath you in the atmosphere to well above the spacecraft's altitude.

FOREWORD BY KATHY SULLIVAN

Astronaut Kathy Sullivan (above). Satellite images (right) simulate a 25-hour view of Earth as the planet would appear from a point in space.

The most captivating sight is lightning. Huge storm complexes, stretching hundreds of miles in every direction, are alive with flashing light. Sometimes the lightning dances randomly from point to point within the storm. Other times there is a rapid barrage in one place. Or one flash will seem to trigger a ripple-fire effect throughout the system.

A typical Space Shuttle orbit lasts for 90 minutes, with a sunrise and a sunset occurring on each revolution. As you near the terminator—the line between the day-lit side and the dark side of Earth—the oblique lighting gives every scene a dramatic texture. Along the ground, clouds cast immense spikes of shadow hundreds of miles long. Topography appears in striking relief; the world grows progressively darker. And all this happens while full, bright sunlight continues to bathe your spacecraft.

Most space fliers say they return from orbit with a very different sense of Earth from what they had before. They have a greater appreciation of our planet and a greater respect. I think space fliers develop a different sense of their relationship with the planet, in somewhat the same way that a person's first Earth-bound travels abroad give new meaning to citizenship and national identity. Perhaps what awakens in orbit is a sort of "planetary patriotism."

I've often wished there were some way to provide everyone with a dose of the curiosity and fascination about Earth that views from orbit inspire. But it will be a long time before orbiting Earth becomes commonplace. So I've had to search for other means of communicating my experience.

Looking at Earth is a superb answer. The space views in these pages are not the personal, human-eye views that I have described. They are the even more important images obtained by the many satellite systems that scientists use to monitor and study our Earth. The imagery is woven into a matrix of scientific, historical, and cultural information. This, I believe, will educate readers and motivate them to develop a deep interest in Earth and environmental sciences. We are likely to face increasingly complex environmental issues and tough decisions over the next several decades. The more informed we are, the better will be our chances to make sound decisions in households, corporate offices, and the halls of government. I know that *Looking at Earth* will give you a sense of seeing what I have seen, a sense that we are not merely inhabitants of this wonderful planet but also its citizens and its stewards.

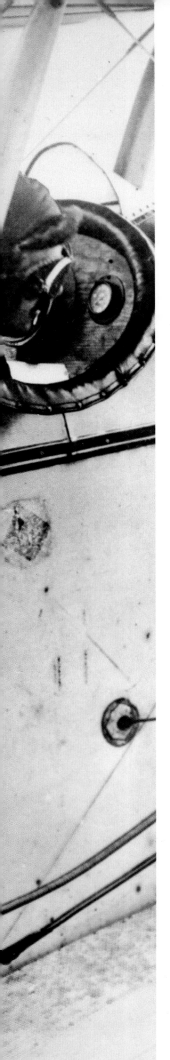

THE EARTH FROM ABOVE

SINCE EARLIEST TIMES, HUMANKIND HAS SOUGHT TO BETTER UNDERSTAND the world by viewing it from above. From a hilltop or the vantage point of a fortress tower, an observer could look at the land and plan for obstacles or dangers that lay ahead. Today travelers on aircraft and spacecraft can look down on Earth with a new perspective. Political boundaries fade, and new patterns emerge to reveal landscapes of beauty and symmetry, undetected from the ground.

In this book we shall look at the Earth continent by continent and observe from space the magnificence and diversity of natural and human-built features that shape the land. Starting with a broad view of an entire continent or region, each chapter then presents scenes that illustrate aspects of the area's terrain. Scenes range from glaciers to deserts, from areas untouched by human activity to crowded urban centers where large populations make a visible impact on the land. The views in this volume represent an international effort of scientists, astronauts, and engineers to monitor and study the Earth.

The science of remote sensing—the gathering of data from a distance—traces back to the early development of telescopes and cameras. By the late 1850s,

Peering down a camera viewfinder from the open cockpit of a Curtiss Jenny, a flier practices the early techniques of aerial photography. Today, aircraft and spacecraft routinely look down on the Earth to survey terrain, predict the weather, monitor crops and forests, plan cities, locate new resources, and gather intelligence.

The rooftops of Boston stretch out toward the harbor in this 1860 bird's-eye view. The steeple of the Old South Church can be seen at left. James Black, a Boston photographer, recorded the scene when he was 370 m (1,200 ft.) over the city in a balloon named "The Queen of the Air." The photograph is believed to be the first clear aerial image taken in the United States.

photographers were flying in balloons to try to obtain aerial photographs. The first to achieve success was the French photographer and publisher Gaspard Félix Tournachon (also known as Nadar), who produced a crude aerial photo of the outskirts of Paris in 1858. Kites and even pigeons (see pages 17-18) were also used for aerial photography in the early days.

With the development of the airplane came an ideal platform for looking down on Earth. Even though a moving, vibrating aircraft produced problems for photographers, the technology of aerial observations evolved rapidly. By World War I, aircraft such as the de Havilland DH-4 biplane were flying photo reconnaissance missions. After the war, the DH-4 flew forest patrols, made geologic reconnaissance flights, and collected aerial photos for mapping and stereo viewing. In fact, for more than a decade following the war, the DH-4 was the standard airplane for aerial photography in the United States.

Dramatically contrasted with the open-cockpit, low-altitude DH-4 is the U-2 high-altitude, reconnaissance aircraft designed some 40 years later. At first the U-2 could fly over the Soviet Union unchallenged by Soviet interceptors and anti-aircraft missiles unable to match its performance. By 1960, however, a major international incident flared when the U-2 flown by Francis Gary Powers was shot down during a reconnaissance mission in Soviet air space.

As the technology of aerial photography grew, more and more practical uses developed: environmental monitoring, disaster assessment, urban planning, geologic mapping. After a severe earthquake, when access to devastated areas may be difficult, photo analysis can help to quickly determine sites most in need of assistance. In 1962, for example, underground coal fires began to burn uncontrollably around Centralia, Pennsylvania. Aerial heat-detecting sensors produced images showing where the ground surface was warmer than normal. The images quickly produced a map of the underground fires and enabled firefighters to proceed more effectively.

Aerial photography also provides an historic record for documenting changes in the development of cities and landscapes and for locating environmental hazards. In 1979, a recreational park was being constructed on Neville Island, Pennsylvania. When workers became ill, investigators speculated about an unrecognized hazardous waste site. Aerial photos of the island as far back as 1938 revealed that solid and liquid wastes had been dumped there for years. Pools and trenches pinpointed on the photos indicated the areas in need of cleanup. In this and many other similar instances aerial photography furnished a broad view that made the detailed work on the ground more efficient.

The next logical step up was space. Early manned and unmanned spacecraft gave a new perspective. Vast areas could be covered by one frame and monitored for seasonal, and even daily, changes. TIROS, the first weather satellite, launched in 1960, started the era of daily weather and storm prediction we now take for granted. Prior to satellite monitoring, the only

way coastal areas received warning of an impending hurricane was if a plane or ship happened to encounter the storm at sea. In 1900, a fierce storm that suddenly struck the coast near Galveston, Texas, killed 12,000 people. Today weather satellites constantly monitor the formation of severe storms and then track them, making it possible to predict their paths.

By the early 1960s, the United States had developed and sent aloft photo reconnaissance satellites known as "eyes in the skies." Kept secret from the public, the satellites' images, along with information from electro-optical and radar reconnaissance, gave U.S. policymakers insights into activities in the Soviet Union and other places of national security interest.

In addition, the early Gemini missions of the same period produced the first successful photographic observations of large-scale geologic structures from space. This success led to the development of the Landsat (land satellite) program. Today information provided by Landsat data aids in crop yield estimates, forest inventory, and rangeland management. Landsat also surveys soil for erosion and moisture, monitors droughts, maps flood plains, explores for oil and minerals, provides data for decisions about regional and urban land use, discerns ocean circulation patterns, and helps environmental scientists studying wildlife habitats and pollution problems.

Five Landsats have been launched to date, the first in July 1972. Landsats 1, 2, and 3 operated at an altitude of 918 km (570 mi.) in a near-polar, north-to-south orbit. This orbit provides repetitive coverage every 18 days. A continuous swath of imagery was acquired along the orbit path and then transmitted electronically to a ground station. These image strips were later subdivided into "scenes" that covered a 185 by 185 km (115 by 115 mi.) area on the ground.

The first three Landsat spacecraft were deployed with two imaging systems that collect data in different parts of the spectrum (see pages 26-29). One system, the Multispectral Scanner (MSS), uses an oscillating mirror that scans the Earth's ·surface from west to east. During these scans, rows of detectors measure the intensity of energy reflected or emitted in different wavelengths from the surface

Stalwart pigeon photographers (above) prepare for work. The tiny pigeon cameras were designed in 1903 by Julius Neubronner, a German photographer. Each camera weighed about 70 g (2.5 oz.). An automatic timer fitted on the camera made an exposure every 30 seconds as the carrier pigeon flew back and forth above the Earth.

Whole city blocks lie in ruin after the disastrous San Francisco earthquake of 1906. G. R. Lawrence, a photographer from Chicago, arrived six weeks after the earthquake and captured this panoramic view of the devastated city from an altitude of about 600 m (2,000 ft.). Lawrence raised his large camera aloft by hooking it to an array of 17 kites.

and encode this information as a series of 0's and 1's (binary digits or bits) to be signaled back to Earth. This process produces what is known as digital imagery.

A digital image is different from a photograph collected by a camera directly on film. (See pages 26-29.) The digital data form separate images of a scene as it is viewed in different wavelengths. Images from three separate wavelength bands can be assigned a primary color (red, green, or blue), and the bands can be overlaid to create a composite image. The colors, however, do not always correspond to the colors that the human eye would see. One reason these "false-color" composite images are useful is because certain band combinations may highlight features and patterns not otherwise apparent. In standard false-color imagery, for example, strong, healthy vegetation appears bright red, and vegetation under stress shows in lighter red or pink tones.

The launch of Landsat 4 on July 16, 1982, marked the beginning of the second generation of Landsat remote-sensing satellites. These carry a more sophisticated

sensor, the Thematic Mapper (TM), which has a higher resolution and more bands to provide extended sensor range in the visible and infrared spectral regions. Power supply and communications problems restricted the operations of Landsat 4 and forced Landsat 5 to be launched earlier than scheduled on March 1, 1984.

Landsats 4 and 5, well past their design lifetime, are operated only intermittently in response to specific data requests. Landsat 6, which will carry an Enhanced Thematic Mapper instrument, is scheduled for launch in early 1993.

Unlike the newer digital systems, film cameras have been aboard spacecraft since the earliest days of the U.S. space program. Mercury and Gemini astronauts carried cameras to photograph the features of the Earth. Apollo and Skylab missions also recorded Earth imagery on film. Hand-held cameras are carried on each Space Shuttle mission. Most Shuttle photography is taken in natural color with a Hasselblad Model 500 EL/M 70mm camera with three different lenses: 50, 100, and 250 mm. Occasionally color infrared film is also used.

During the Civil War, Thaddeus Lowe (below, in basket) pioneered the use of balloons for gathering military intelligence. More than 70 years later, the high-altitude balloon Explorer II (opposite page) took the first photograph recording the curvature of the Earth.

On a Shuttle mission in late 1983 the European Space Agency's Spacelab pod flew in the Shuttle cargo bay. After the Shuttle achieved orbit, an astronaut mounted Germany's Metric Camera in the Spacelab's window. The camera made more than 900 color infrared and black-and-white, high-resolution photographs.

The former Soviet Union has begun to market images acquired from manned and unmanned spacecraft. Exposed film from unmanned missions is returned to Earth in recoverable pods. Some of this photography has a spatial resolution of 5 to 8 m (16 to 26 ft.), meaning that it can discern objects down to that size. This is the best spatial resolution of any Earth imagery currently commercially available. (For information on other cameras, see pages 298-301.)

As digital spacecraft imagery of the Earth from space proved itself useful to scientists in a variety of fields, several nations joined the effort to monitor the Earth's environment from orbit. The first French SPOT satellite was launched on an Ariane rocket in early 1986. SPOT is the first satellite remote-sensing program to be designed and operated as a commercial system. It provided digital black-and-white and multi-spectral imagery through the end of 1990. SPOT 2 began operation in January 1990 and took over all imaging duties after SPOT 1 was retired. SPOT 1 was reactivated in March 1992 so that both satellites could work together to meet imaging needs. Although the SPOT satellites operate in fewer spectral bands than the Landsat sensors, they provide a better spatial resolution.

In 1987 the Japanese launched their first Earth observations mission MOS-1A (Marine Observation Satellite) from the Tanegashima Space Center on an island off the southern coast of Japan. MOS-1A, now joined by MOS-1B, was designed to monitor coastal areas and provide multi-spectral imagery with a resolution of 50 m (160 ft.). Its instruments measure the temperature and color of the sea surface and can be used for the study of currents, and suspended sediments.

The Indian Remote Sensing Satellite, IRS-1A, began operating in 1988. It was joined by the IRS-1B in August 1991. Both were launched aboard Vostok rockets from Khazhakstan in the former Soviet Union. Their coverage is limited to India and nearby areas. The satellites were designed for environmental studies and to help monitor forests, agricultural development, and urban growth.

Satellites like the Landsat, SPOT, MOS, and IRS are considered passive remote-sensing devices because they measure energy naturally reflected or radiated from the Earth. Radar is an active remote-sensing system because it provides its own signal, which is bounced off a target, such as an aircraft or a mountain, and returned to a receiver. Radar systems, therefore, "illuminate" their targets themselves and do not rely on sunlight. They can operate day or night and, because radar penetrates clouds, in any weather. This makes radar especially useful in tropical areas where clouds nearly always block the view of passive sensors.

Radar systems have also proved useful in arid regions. Radar imagery of

desert lands has located ancient waterways now dry and obscured by sand. Radar has also been used to map the paths of old caravan trails that led to the remains of a long-lost city. It may be the fabled trading center of Ubar, which flourished in Oman more than 4,000 years ago.

Several orbiting radar systems have been launched to study various aspects of land and sea. Seasat, which operated briefly in 1978, was used to study ocean phenomena, including sea-surface roughness, ocean currents, icebergs, oil spills, and coastal features. It orbited at an altitude of 790 km (490 mi.) and produced an image swath 100 km (62 mi.) wide.

Space Shuttle missions flew two radar instruments: Shuttle Imaging Radar-A (SIR-A) in November 1981 and SIR-B in October 1984. These radar experiments were stowed inside the Shuttle cargo bay and operated when the Shuttle was in an inverted position (open cargo bay facing the Earth). Shuttle imaging radar missions acquired image swaths 20 to 50 km (10 to 30 mi.) wide and were primarily flown for terrain observations and geological studies.

An international array of new radar sensors has been orbited in the last few

U-2 aircraft like this one (right) gathered intelligence as in the photograph (left) revealing Soviet submarines in the Barents Sea. Photographs taken from U-2s helped confirm the Soviet missile buildup in Cuba in 1962, verified nuclear weapon tests in China, and provided tactical military intelligence during the Vietnam War and the Persian Gulf War. U-2s have also monitored Earth's resources and atmosphere.

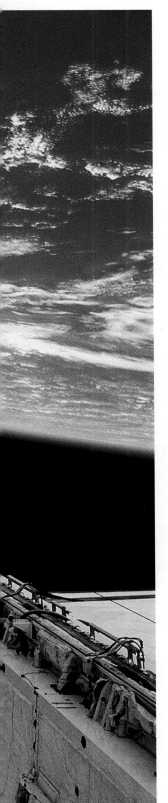

years. In March 1991, the former Soviet Union launched the Almaz satellite, which is designed to provide high-resolution (15 m or 50 ft.) radar imagery. The European Space Agency's European Remote Sensing Satellite (ERS-1), launched in July 1991, includes a radar instrument mainly targeted for ocean studies. The Earth Resources Satellite of Japan (ERS-1 or JERS-1) joined the group in early 1992. It carries both a multi-spectral scanner and a radar sensor for high-resolution mapping. These systems should provide greater and more detailed coverage of the Earth's surface than ever before.

Today the Earth is constantly monitored by a variety of spacecraft. They record Earth's structure, its changes, its catastrophes, and its beauty. This view from above contributes to both understanding and appreciating our world.

Throughout the years, humankind has traveled continually higher and farther away to explore the beyond and to pursue many different goals and challenges. Yet, to many who have participated in these exciting voyages, the best part of all was looking back at Earth, the planet we call home.

As a satellite is released from the cargo bay, the Space Shuttle glides above the blue oceans of Earth (left). A more distant view from the Voyager 1 spacecraft (right) reveals both the Earth and Moon in the void of space. This image was recorded in 1977, when the Voyager was 12 million km (7.5 million mi.) from Earth.

The Electromagnetic Spectrum

VISIBLE LIGHT IS MADE UP OF A SPECTRUM OF COLORS: RED, ORANGE, YELLOW, Green, Blue, Violet. The colors represent a range of different wavelengths of light, or electromagnetic radiation, with longer wavelengths near the red end of the spectrum and shorter wavelengths near the violet. Objects reflect and absorb different wavelengths of light, and the human eye perceives these differences as color. For example, an object the eye sees as red reflects red light and absorbs much of the other wavelengths. When electromagnetic radiation occurs with wavelengths longer than those of red light (i.e. infrared) or shorter than those of violet light (i.e. ultraviolet), it is beyond the capability of the eye to see.

The visible wavelengths comprise a very limited area of the spectrum. However, the remote sensors on spacecraft can detect wavelengths beyond the visible range. These include naturally emitted thermal infrared radiation and radar, both of which can be measured at night. A thermal infrared image resembles a temperature map of the Earth's surface. Unlike sensors that detect visible light, radar instruments see right through rain, snow, and clouds.

Sampling the Spectrum

Three views of the Salton Sea region in southern California are shown. The photo on the bottom represents the visible range. The green vegetation of the cultivated fields contrasts sharply with the narrow strip of sand that makes up the Algodones dune field. The photo in the center was taken on film sensitive to infrared (IR). The vegetation is emphasized and appears red. On the top is a radar image of the scene collected by the Seasat satellite in 1978. Unlike the two previous images, which were made by a camera passively recording the radiation reflected by the terrain below, the radar scene was constructed through an active process whereby a signal was sent to the target and bounced back. The smooth surface of the dune field appears dark while the rough texture and orientation of the mountains cause them to return a stronger radar signal and therefore appear bright.

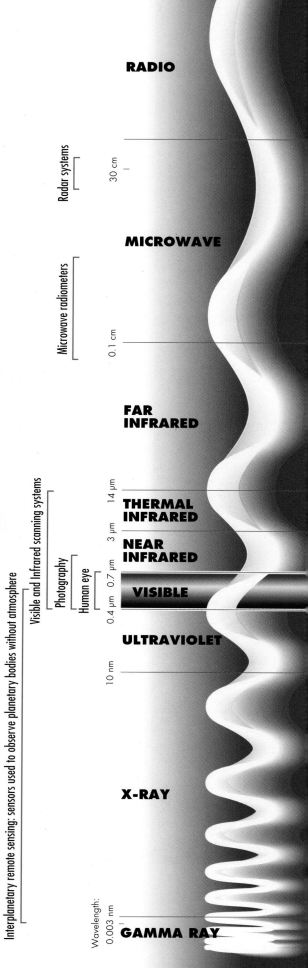

RADIO

Radar systems

30 cm

MICROWAVE

Microwave radiometers

0.1 cm

FAR INFRARED

Visible and Infrared scanning systems

14 µm

THERMAL INFRARED

3 µm

NEAR INFRARED

Photography

Human eye

0.7 µm

VISIBLE

0.4 µm

ULTRAVIOLET

10 nm

X-RAY

Interplanetary remote sensing: sensors used to observe planetary bodies without atmosphere

Wavelength: 0.003 nm

GAMMA RAY

NUMERICAL VALUES

101	90	93	101	60	52	78	101	123	120	97	97	90	93	108
150	135	82	78	142	123	146	153	131	116	120	90	67	71	
165	172	82	101	176	187	165	180	206	216	176	165	121	105	142
198	202	123	135	206	198	131	93	108	168	191	157	135	207	243
183	210	202	195	202	165	82	67	108	180	213	172	187	247	228
138	191	210	221	195	127	78	157	243	251	221	165	150	127	116
183	195	168	202	198	150	67	90	213	243	191	142	90	71	101
225	221	191	206	243	210	78	0	127	255	198	142	172	120	71
228	213	202	228	236	202	157	131	228	255	195	131	168	150	82
210	161	150	183	172	161	191	213	255	255	172	105	142	120	97
176	165	153	176	183	142	120	108	180	236	198	105	105	183	180
157	210	198	168	198	161	138	187	243	255	255	127	180	255	255
161	157	131	108	165	198	255	255	228	131	150	206	217	127	
138	116	97	112	157	213	243	243	165	123	183	172	101	97	116
161	93	82	93	123	180	195	150	93	86	127	161	142	138	108
105	90	90	105	131	120	93	108	97	86	116	127	101	105	127

75	81	98	104	92	86	98	115	121	133	121	139	104	69	86
104	98	81	86	104	98	81	98	133	179	162	92	81	69	75
139	110	86	86	115	98	104	156	231	231	156	98	81	92	75
139	104	98	110	115	110	104	98	133	208	243	162	75	92	115
115	104	121	156	115	98	133	185	224	255	255	162	104	104	98
121	115	110	121	133	104	173	255	255	255	179	110	81	86	86
115	127	133	168	208	156	162	191	255	243	191	179	144	92	86
98	110	121	220	255	255	173	255	173	185	226	133	75	81	
104	92	127	237	255	255	255	255	255	202	191	202	173	75	75
127	110	121	197	226	231	249	255	255	156	156	156	179	121	92
150	133	115	150	156	168	220	255	255	214	144	98	110	98	110
115	115	98	98	110	127	237	255	255	255	168	104	110	92	104
98	121	127	127	144	255	255	255	255	173	92	92	81	81	
133	92	98	104	86	110	156	173	150	179	214	173	98	81	81
104	86	86	92	81	69	63	104	168	133	121	110	92	92	
75	81	92	86	98	98	81	104	98	110	144	115	92	110	104

91	97	91	109	109	99	97	109	127	139	139	127	109	72	66
109	115	78	97	109	91	91	91	133	188	176	103	91	109	109
145	139	78	85	97	97	109	109	151	230	242	163	115	103	103
139	145	121	109	121	121	115	133	206	236	194	109	103	97	103
121	127	115	127	145	121	85	182	255	255	255	162	133	103	103
127	115	127	188	188	200	176	255	255	255	200	188	151	109	72
121	115	121	212	255	255	188	206	255	255	182	224	224	139	78
115	103	121	212	255	255	255	255	255	188	188	212	170	97	91
109	103	115	176	212	224	248	255	255	145	145	163	163	115	103
139	145	133	139	127	206	255	255	224	151	103	115	115	127	
121	133	103	109	115	115	255	255	255	255	145	97	103	97	115
115	97	127	188	188	200	176	255	255	255	170	121	91	85	91
133	97	97	109	109	127	163	170	145	194	212	170	109	97	97
121	109	91	91	109	115	91	85	151	151	145	133	109	97	91
121	109	97	91	97	103	91	91	109	103	109	145	121	103	115
85	97	97	97	103	91	91	109	103	109	145	121	103	115	97

BRIGHTNESS LEVELS

COLOR ASSIGNMENTS

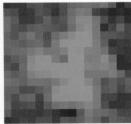

COLOR COMPOSITE

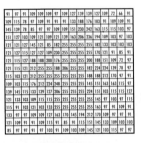

Digital images

Digital images are collected as a series of numerical values, each value representing the amount of energy radiated or reflected from a unit of area on the Earth's surface. Digital images have some advantages over photographs. In satellite remote sensing, digital sensors are flown on longer-duration platforms while cameras are usually limited to shorter missions, giving systems such as Landsat greater repetitive coverage than film systems. Digital data are more readily analyzed by computer processing. Digital images can be enhanced by processing, improving the contrast and exaggerating features of interest. Perhaps most importantly, digital images are collected in separate wavelength bands including, and beyond, the region of the spectrum detectable by film systems.

Information from each band produces a discrete image from the swath of Earth below the spacecraft. Each image is composed of pixels (picture elements) which can be compared to the squares on a checkerboard. Individual pixel values (numbers representing the relative brightness of each point) are transmitted to a receiving station on Earth to produce the rows and columns of a numerical matrix comprising each scene. Pixels with high values will appear bright. Those with low values will be dark.

If a picture is worth a thousand words, a color composite image is worth many more because three images are combined into one. This is possible because the images were recorded simultaneously from the same vantage point and because they are digital. Each of the three images of the scene, measured in different wavelengths, is assigned a different primary color: red, green, or blue. The brighter the pixel the more intense the color. When corresponding pixels (the same row and column) from each image are added together, the resulting color is a hue that represents the proportion of red, green, or blue from each of the three original digital images.

The large image (left) shows a Landsat Thematic Mapper scene of Washington, D.C., including the Potomac and Anacostia rivers. The box indicates the area around the U.S. Capitol building that is enlarged below.

Digital images of the Capitol area of Washington, D.C., from three different spectral bands are combined into a "false-color" composite. The colors are called "false" because any primary color can be assigned to any band. Thus, vegetation can be featured in red by assigning this color to a near infrared band. Vegetation is very reflective in the near infrared and, therefore, has high brightness values in this band.

View From a Distance

SOLAR PANELS | SENSOR | ANTENNA

MONITORING THE EARTH FROM DISTANCES GREATER THAN 200 kilometers above the ground has become an international effort in the two decades since the United States launched the first Landsat satellite. The photographs and digital images in this book are from some of the most advanced remote sensors from the U.S., France, Japan, Germany, and the former Soviet Union. These remote sensing systems forewarn us about impending disasters from floods, hurricanes, toxic Red Tide plankton blooms, and volcanic plumes. Longer-term disasters, such as famine, ozone hole formation, forest die-back due to acid rain deposition, and habitat loss through changing land use patterns are also monitored from space. High-resolution reconnaissance satellites are used to detect arms buildups and to verify compliance with nuclear disarmament treaties. Unmanned spacecraft provide continuous and repetitive coverage that is useful for monitoring changes through time. Astronauts, however, have the ability to photograph unexpected transient events, such as storms and volcanic eruptions.

Collecting Images from Space

As satellites orbit the Earth, their digital sensors continually monitor the scene below at predetermined wavelengths of the spectrum (bands). The digital images are stored on computer-compatible tapes and can be processed in a variety of band combinations to accentuate different aspects of the terrain.

Other spacecraft, such as the Space Shuttle, carry a variety of film systems. Photographs from space recorded on film using handheld or bracket-mounted cameras are convincing reproductions of what an astronaut's eyes see in the visible wavelengths of the spectrum. Using film sensitive to infrared wavelengths, a camera can also record features that human eyes cannot detect.

Spatial Resolution

Resolution is a measure of how detailed an image is. The most direct way to determine the spatial resolution of an image is to find the minimum ground distance at which two objects in the image can be distinguished as separate entities. That is, if the two objects were any closer, they would appear to be a single feature. This test of ground resolution does not require recognition or identification of the objects, merely the ability to detect them as separate features.

The spatial resolution of photographs taken from spacecraft varies widely with the altitude of the spacecraft and the focal length (lens size) of the camera. Very generally, most Space Shuttle Hasselblad photographs acquired with a 100-mm lens have a resolution similar to Landsat MSS, while using a 250-mm lens produces a photograph with a resolution similar to Landsat TM. The resolution of digital scanning sensors, such as Landsat MSS and TM, usually remains constant throughout the system's operational life because the spacecraft's altitude changes little during this time. The table (above) compares the resolutions of some of the digital remote sensors included in this book.

30 M RESOLUTION

120 M RESOLUTION

Remote sensors	Nominal spatial resolution
Advanced Very High Resolution Radiometer (U.S.)	1.1 km
Heat Capacity Mapping Mission (U.S.)	600 m
Marine Observation Satellite (Japan)	50 m
Shuttle Imaging Radar A & B (U.S.)	40, 20 m
Landsat Multispectral Scanner (U.S.)	80 m
Landsat Thematic Mapper (Visible and near IR bands)	30 m
Seasat Radar (U.S.)	25 m
SPOT panchromatic & multispectral (France)	10, 20 m

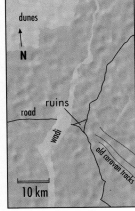

Archaeology from Space

The discovery of a lost city is an archaeologist's dream, but decades have been spent fruitlessly searching huge deserts or thick jungles for legendary sites. With remote sensing, large areas can be quickly scanned for clues.

One example of this application is the recent discovery of an ancient city in southern Oman. The site was located using images acquired from space, including Space Shuttle Radar, Landsat (left), and SPOT. The images revealed converging tracks of old caravan routes, not visible from the ground or with conventional photography. A team of archaeologists went to Oman and found ruins where the tracks converged.

The ruins are thought to be Ubar, a lost city described by T. E. Lawrence (Lawrence of Arabia) as "the Atlantis of the sands." Archaeologists (below) excavate to confirm the existence of the city.

A MASSIVE LAND

A JOURNEY NORTH OR SOUTH FROM THE INTERIOR OF EQUATORIAL Africa takes a traveler from lush, humid tropical forests to open woodlands, then through grasslands, and finally across arid deserts. From these varied country-sides the homelands of many cultures and many peoples have developed: Africa is home to desert nomads and river fishermen, to floodplain farmers and city dwellers. Their land supports such cultural diversity that more than 800 different languages can be heard on this continent.

Africa's unusual variety of landscapes and climates is primarily a result of the continent's large area and unique location. Africa, second to Eurasia in landmass, is the only continent centered on the equator. Africa's equatorial position creates a somewhat symmetrical continent, with similar climate and vegetation zones in both the north and the south. This makes Africa much different from the other equatorial continent, South America, where the equator lies far to the north of the continent's center.

The diversity of the African landscape can be seen in the continental mosaic of satellite images on the opposite page. The false-color images show the sands of the Sahara Desert (light brown to white) spreading across much of northern Africa. The images also reveal that the desert is not simply a vast sea of flat-lying sands and dunes. The Sahara contains large areas of gravel and exposed rock (gray to black areas). Some of the exposed rock in the Sahara has volcanic origins, such as the Tibesti Massif in northern Chad (top center), which appears very dark. In false-color, red represents healthy plants. The lack of red in the north, therefore, shows that the desert is sparsely vegetated. The northwestern coast of Africa, along the Atlantic Ocean and Mediterranean Sea, appears red because it receives enough rainfall to support natural vegetation and farming.

The dark gray-red band south of the Sahara is the Sahel, a semiarid grassland that provides an austere homeland to pastoral nomads. In the past, the nomads, because of their mobility and small population, had little impact on the land. Beginning in the 1960s, however, West Africa's newly independent nations created political borders that restricted the nomads' movement from one traditional grazing land to another. At the same time, many of the Sahel's nomads settled on government-owned rangelands.

When they became a more settled people, the former nomads began raising their livestock not for subsistence living but for commercial production. As the herds became larger and competed for grazing land, pressure on the grasslands increased. The human population also grew steadily, making new demands on the land. People farmed intensively and stripped scant woodland for fuel. Then,

NOAA AVHRR MOSAIC

COURTESY OF THE ENVIRONMENTAL

RESEARCH INSTITUTE OF MICHIGAN AND

THE NATIONAL GEOGRAPHIC SOCIETY

starting in the late 1960s, a series of devastating droughts parched the Sahel, further degrading the land.

South of tropical Africa are other regions with semiarid and arid climates. The Okavango Delta is revealed by its distinctive dark, bird-foot shape (lower center). To the southeast of the delta, the Makgadikgadi Ntwetwe salt pan appears blue-gray. The delta and salt pan lie at the northern edge of the Kalahari Desert. Though classified as a desert, most of the Kalahari is covered by vegetation that has adapted to the region's arid climate. The vegetation includes baobab, palm trees, scrub acacia, and grasses and shrubs. In the mosaic this vegetation gives the Kalahari Desert a vastly different appearance from the stark Sahara.

Desert regions are easily viewed from space because they are usually free of clouds. The lack of light-scattering water vapor in arid regions makes it possible to obtain extremely clear images of deserts, if dust storms are not obscuring the sensor's view. Because deserts are not masked by vegetation, their landforms are revealed in fine detail. Africa's deserts contain many fascinating features, such as fields of complex sand dunes, dried channels of long-dead rivers, and wind-sculpted outcrops of desert rock.

Here, too, can be seen the never-ending struggle of life in a hostile, arid land. In some places, such as the Kharga Oasis in Egypt's Western Desert, advancing dunes threaten several villages. The struggle takes place on a broader front in the Sahel, where life's hold along the margin of the world's greatest desert has always been tenuous.

Africa's desert sands and the continent's other surface features, such as swamp-filled basins and tropical forests, overlie an ancient, relatively flat plateau. Geologically, the core of the continent is composed of several stable shields of extremely old crystalline rock. Africa, unlike other continents, has only narrow coastal plains and continental shelves, and lacks long chains of mountain ranges. The absence of such mountain ranges suggests that Africa has been spared much of the geologic upheaval experienced by other continents. There are relatively few earthquakes on this continent.

The eastern part of Africa, however, shows the results of momentous geologic forces that have created rift valleys. These rifts form when the Earth's crust is fractured and the two sides of the fracture spread apart.

Of all the continents, Africa has the most extensive, well-defined system of rift valleys. In East Africa, these valleys can be traced by lakes, generally oriented north-south, that bracket Lake Victoria (center right).

The island of Madagascar (mosaic, lower right) and the coastal portions of west-central Africa are veiled by cumulus clouds. Because cloud cover and heavy rains are typical of tropical regions, clear satellite images of the tropics are rare.

Many of the continent's major rivers originate in tropical or subtropical Africa and reach the sea only after following lengthy, circuitous courses. The longest river on Earth, the Nile (mosaic, upper right), flows to the sea through a swamp-filled basin in southern Sudan and across the largest desert in the world. The other great rivers of Africa, with their floodplains, swamps, and channels, are also clearly visible from space. Along these rivers' courses the satellite images show signs of human activity, including dams, canals, and irrigated farmland.

This map (right) shows the location of the images in this chapter. Red lines indicate the geographic coverage of each scene. The different sizes of each frame reflect the range of altitudes and capabilities of a variety of sensors.

MEDITERRANEAN SEA

ITALY
GREECE
TURKEY
CYPRUS
SYRIA
LEBANON
ISRAEL
IRAQ
IRAN
JORDAN
KUWAIT
SAUDI ARABIA
YEMEN

TUNISIA
ALGERIA
LIBYA
EGYPT
NIGER
CHAD
MALI
TOMBOUCTOU

AFRICA

SUDAN
ETHIOPIA
SOMALIA
DJIBOUTI

no defined boundaries

RED SEA

Nile R.
L. Nasser
Blue Nile R.
White Nile R.

BURKINA FASO
GHANA
TOGO
BENIN
NIGERIA
CAMEROON
CENTRAL AFRICAN REPUBLIC
Lake Chad
Niger R.

EQUATORIAL GUINEA
GABON
CONGO
ZAIRE
Zaire R.

UGANDA
KENYA
RWANDA
BURUNDI
Lake Victoria
Lake Turkana
TANZANIA
Lake Tanganyika

ANGOLA
ZAMBIA
Zambezi R.
Lake Malawi
MALAWI
ZIMBABWE
BOSTWANA
NAMIBIA
MOZAMBIQUE
MADAGASCAR

SOUTH ATLANTIC OCEAN

SWAZILAND
LESOTHO
SOUTH AFRICA

INDIAN OCEAN

p.44 Tifernine Dunes
p.42-43 Kharga Depression
p.51 The Nile
p.52 Aswan Dam
p.52-53 Lake Nasser
p.44 Vanished Rivers
0-41 Tombouctou
p.41 Tombouctou
p.37 Djibouti
p.36 Lake Turkana
p.57 Zaire River
p.35 Land of Volcanoes
p.45 Visible border
p.58-59 Madagascar
p.56 Okavango Delta
p.46-47 Namib Desert
p.47-49 Sossus Vlei

The Great Rift

MUCH OF THE AFRICAN CONTINENT RECORDS A HISTORY of relative geological stability. But the eastern edge, characterized by Africa's rift valleys, is being torn apart by movement in the Earth's crust. Numerous volcanoes, both active and extinct, occur throughout the rift region. In width, the rift valleys range from about 30 km to more than 200 km (20 to 120 mi.). The East African Rift extends through Ethiopia southward to Mozambique. About halfway along its course, the system divides into east and west branches on either side of Lake Victoria. The rift system is part of a longer structure stretching up through the Red Sea to Lebanon and Turkey.

The entire East African Rift cannot be seen in a single glance from ground level. A mosaic of satellite imagery, however, can produce a view of the rift in its entirety and give scientists a better understanding of its structure.

Africa's great rift formed as blocks of the Earth dropped between parallel faults. This rifting is related to the motion of large sections of the Earth's crust, called plates. The theory of continental drift describes how the present continents broke off from a single large landmass, named Pangea, about 200 million years ago. The continents drifted across the globe by spreading apart at rifts and mid-ocean ridges. The matching outlines of western Africa and eastern South America provide evidence of an ancient split that occurred when the South American plate broke away from Africa.

Inevitably, plates collided, either slipping past each other in huge fault zones like California's San Andreas, plunging beneath one another to form belts of volcanic activity and earthquakes, or hitting head-on to buckle and fold the terrain into great mountain ranges.

Rifts along the Red Sea and the Gulf of Aden separate the African plate from the Arabian plate. The East African Rift forms a third branch, which intersects these two fault systems in the Afar region of northeast Ethiopia and Djibouti (see page 37). Motion across these rifts is now separating Arabia from Africa and slowly moving East Africa apart from the rest of the continent, although the rate of movement is only a few millimeters per year.

Scarps and linear fault structures mark the fractured terrain of the African rift as it passes through southern Kenya and northern Tanzania (opposite page). The scene illustrates the dramatic geology of the Rift Valley with its lakes, steep scarps, and volcanic structures.

Many of the lakes that follow the valley's course are highly saline, but Lake Baringo, with its bright-blue color on the satellite image, is a freshwater body. Baringo's central island is a small extinct volcano. The lake area is known for its hundreds of species of birds. Heart-shaped Lake Naivasha, the only other freshwater lake in this part of the Rift Valley, harbors a bird sanctuary on its arc-shaped Crescent Island, the remnant of an ancient volcanic crater rim protruding above the lake surface.

Geologists predict that the Great Rift Valley will eventually broaden into a sea, where whales will replace the Masai cattle that now graze in the semidesert of Kenya. The walls of the rift border one of the soda lakes, Little Magadi (above), found in the valley. Without any drainage outlet, evaporation leaves these lakes highly concentrated with salt.

Lake Natron, farther to the south, is a highly saline lake where few fish can survive, but hundreds of thousands of flamingoes often gather there to feed on algae that grow in the brine. The major source of the Natron salts is the neighboring Lengai volcano, a geologic oddity that erupts materials rich in sodium. What appears to be snow, the bright spot at its summit, is actually a deposit of sodium carbonate. The volcano is active, forming volcanic features on the floor of its crater. Lava flows were reported as recently as 1991.

At the upper right is Mount Kenya, an extinct volcano nearly 5200 m (17,000 ft.) high. The white patch on the mountain's peak marks its permanent cover of snow and ice. The red color on its flanks indicates vegetation.

Land of Volcanoes

East of the Rift Valley is the highest mountain in Africa, Mount Kilimanjaro. Kibo, its tallest peak, is nearly 5960 m (about 19,350 ft.) high. This huge glacier-capped mountain is a dormant volcano, with numerous smaller cones on its flanks.

Ngorongoro Crater is a large caldera atop another inactive volcano. The crater has become a magnificent game park and is home to a wide variety of wildlife, including zebras, gazelles, elephants, and lions thriving in the grasslands that have taken root there.

Ancient layers of ash erupted from Ngorongoro are found to the west in Olduvai Gorge, the site of early fossil remains of what many believe to be human ancestors. The gorge lies on the edge of the Serengeti Plain, a fertile grassland known for the great herds that migrate across it every year.

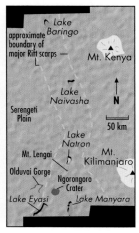

Lake Turkana

Its narrow shape controlled by faults of the Great Rift, Lake Turkana touches the border between Ethiopia and Kenya. Turkana is fed mainly by the Omo River, which creates a swampy delta where it enters the lake from the north. The muddy brown waters near the Omo Delta show the heavy load of sediment the river deposits on the lake bottom. Lake Turkana lies in a semiarid region, where water rapidly evaporates. The evaporation concentrates salts and makes the lake's waters brackish.

Several young volcanoes form islands in the lake. Central Island, a dormant volcanic cone, has become a breeding ground for crocodiles. Off the lake's northeast shore is Koobi Fora, a rich archaeological site where fossils reveal the long history of human habitation around the lake's fish-filled waters.

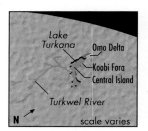

Djibouti

The Republic of Djibouti lies in the Afar region of northeast Africa, a triangular-shaped lowland of searing heat and desolate terrain where the East African Rift joins with two rifts that run through the Red Sea and Gulf of Aden. The capital, a port city also named Djibouti, is just to the east of this photo (above), along the coast of the Gulf of Tadjoura, an arm of the Gulf of Aden. The former French colony is a land of lava flows and lakes.

Lake Assal (above, center), whose white salt deposits indicate extensive evaporation, lies more than 150 m (490 ft.) below sea level and is the lowest point in Africa. The photo at left shows volcanic cones around the lake's shores. The linear fault structures (above, center and top right) and numerous earthquakes in the region are due to the stresses caused by proximity to the rift junction.

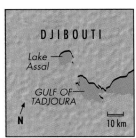

Where Rain Rarely Falls

AFRICA HAS THREE MAJOR DESERTS: THE SAHARA, THE Kalahari, and the Namib. While similar in their aridity, each has a distinct character, which is reflected in its landforms and inhabitants. The Sahara is broad and diverse. Here are ancient volcanic mountains, great sand seas called ergs, and the fabled desert oases, where groundwater makes its way to the surface. The Sahara once extended even farther south than it does today, as shown by the massive ancient dunes that blanket much of the land to the south. To the north, the desiccated landscape of today's Sahara also shows evidence of change: now-dry stream valleys clearly show that there have been times in the not-so-distant past when water flowed abundantly here. From about 6,000 to 13,000 years ago, the climate was wetter and cooler. Early game hunters roamed the waterways, later followed by sendentary farmers and then finally by cattle-raising nomadic peoples who lived along the rivers. Ancient cave and rock paintings in the remote Saharan mountains of Tassili-N-Ajjer provide a dramatic record of climate change. Early paintings are rich with wildlife, while later ones show domestic animals and people gathering grain. Later paintings illustrate camels and the nomadic life similar to the Sahara of today.

As the climate deteriorated, many of these peoples were forced southward into the desert fringe called the Sahel, where there is some water in meager quantities. The rivers that intertwine with the Sahel's ancient sand dunes also are a reminder of the region's history of changing climate.

The landscapes of the Kalahari and Namib also preserve a record of past climate change. The Kalahari is covered with ancient dunes and crossed by now-dry river valleys. Rain, however, does fall; the dunes in some parts of this desert are grass-covered, and the Kalahari supports many kinds of wildlife. The Namib Desert (see pages 46 and 47), just to the west, is a coastal

In much of the Sahel, stumps of trees are all that remain of the vegetated area cleared by fire. In order to plant rice, a Sahelian (above) sacrifices scarce trees.

desert whose dryness is reinforced by the cold Benguela Current flowing just offshore: the cold ocean water causes airborne moisture to condense and fall back into the sea, so that rain rarely reaches the land. Dunes appear mostly in the southern part of the Namib, but the desert itself extends northward into Angola. In the south, the red dunes rise more than 300 m (980 ft.) above the desert floor, and old river valleys make their way from the inland highlands to the coast.

Some of the old rivers still run to the sea when rain falls. One of these is the Kuiseb River, which forms the northern boundary of the dunes. Other ancient rivers, now blocked by the dunes, act as long linear oases amidst the sand. Although the Namib is one of the most arid places on Earth, these linear oases and the cool fogs that rise from the coast enable this desert to support a rich and unique array of plant and animal life.

The ages of the Namib, the Kalahari, the Sahara, and the Sahel are still a mystery, but their landforms are not. All attest to the variability of climate. Drought shapes the Sahel's land and the life of all arid regions, occurring frequently and causing physical, ecological, and social changes. Rivers and lakes become dry, vegetation dies and is not replenished, animal habitats are harmed, and bone-dry soil is further damaged by wind and water. People find it increasingly difficult to make a living or even to survive.

Human reactions to drought accelerate the effects of drought in arid regions. The result sometimes is the ecological disaster called "desertification." Many of the changes that occur during drought and desertification are nothing new for these arid zones. But, as is true for the rain forests and all of the Earth's ecosystems, the speed with which such changes occur and the extent of their impact on the continued habitability of the Earth's arid lands are, at least in part, within humankind's ability to control. (P.A.J.)

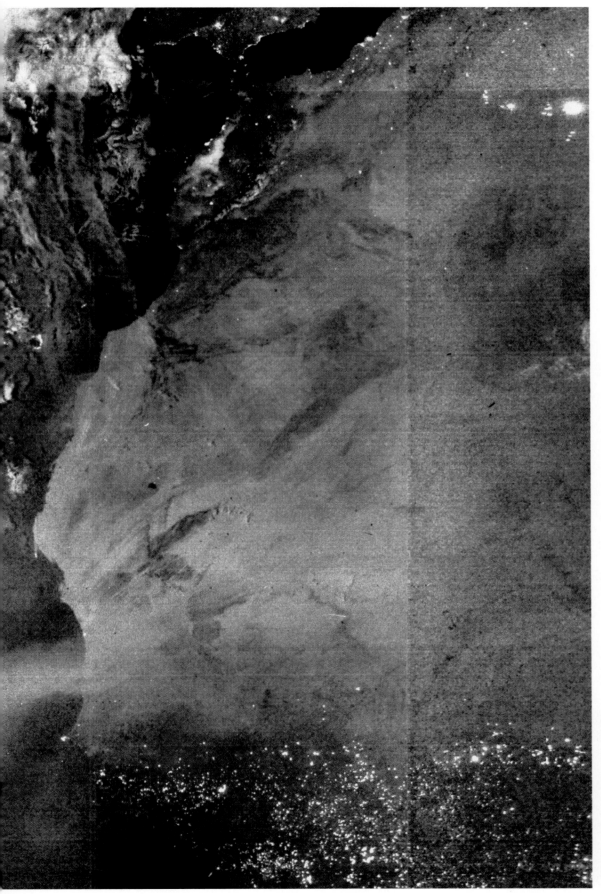

Fires in the Night

In this night view of northwest Africa, fires blaze in a band of bright lights just south of the Sahel. Most are for cooking or to provide warmth and light; others are set to clear land for farming. The array of the fires reflects the density of the population in the region. By contrast, the unpopulated Sahara is dark.

Other bright lights in this image include the port of Dakar in Senegal, cities in Spain, and flares from natural gas burn-off in the oil fields of North Africa. Off the western coast, a dust storm is reflected in moonlight.

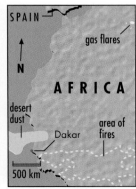

Tombouctou, 1976

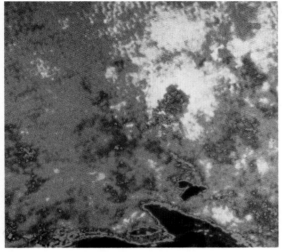

Tombouctou, 1985

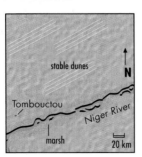

stable dunes

N

Tombouctou

Niger River

marsh

20 km

Tombouctou

A whitish ring, indicating sparsely vegetated savanna, surrounds Tombouctou (Timbuktu) in the lower left of this image (opposite page). For more than a hundred years people on the ground have noticed this ring, now a target for satellite imagery. The ring starkly outlines land denuded by grazing animals and people collecting wood for fuel.

Tombouctou lies in the country of Mali, amidst a field of stable dunes, north of the Niger River. Near the end of the 12th Century, Tombouctou ("Bouctou's well") was a watering post in the care of a woman named Bouctou. The city became a major trade center on north-south caravan routes that crossed the Sahara Desert to the Mediterranean Sea, the Sudan, and the Gulf of Guinea.

At the same time, the city's reputation grew as a center of Moslem teachings. The University of Sankoré was known throughout the Arab world as a center of West African Mohammedan culture.

Camel caravans traveled south out of the Sahara Desert laden with slabs of salt. Ivory, ebony, slaves, and gold were traded along the Niger River and in Tombouctou. The importance of the trans-Sahara trading routes declined when European traders, and later colonizers, arrived by sea on the west African coast.

Tombouctou's decrease in population—from about 100,000 in the 16th Century to 20,000 today—reflects the decline of Saharan trade. By 1984 only two salt caravans per year were arriving in Tombouctou.

Severe, persistent drought struck the Sahel in 1968. In the winter of 1976, the region around Tombouctou still appeared verdant, as shown by the computer-enhanced image (top). The dark-brown area in the center of the image represents Tombouctou itself, and the green color indicates vegetation. That winter, the Niger River floodplain remained inundated well into the dry season, and vegetation lingered on the dunes. The old bright halo around the city was distinct by contrast with its grassy surroundings, but gardens outside the city were green, and the small lakes and channels (blue color) south of Tombouctou were full of water.

After a brief respite, drought returned to Tombouctou, and its environment changed dramatically. In the image (bottom) from the winter of 1985, much had changed. The Niger River floodplain was dry, as were lakes and channels. The vegetation that had formerly covered the dunes was all but gone, leaving the entire region as dry—and therefore as bright—as the formerly narrow halo around the city. (P.A.J.)

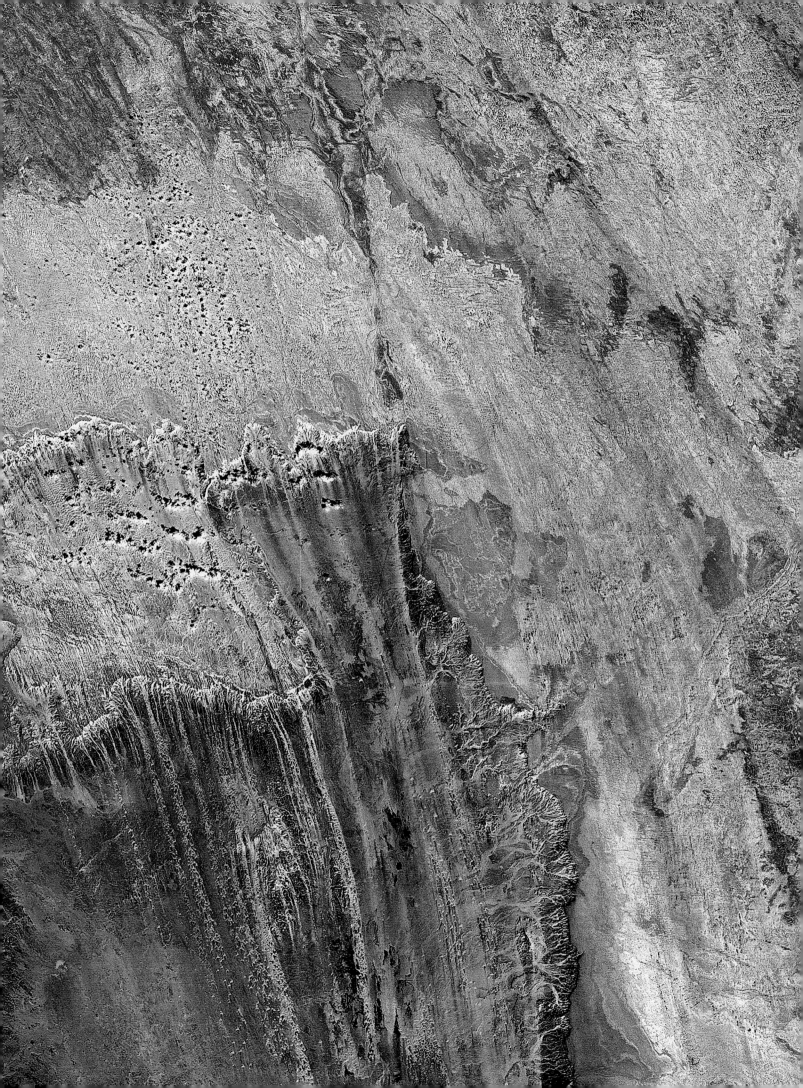

Kharga Depression

In the Western Desert of Egypt is the Kharga Depression, a narrow lowland bordered on the north and east by steep scarps about 200–300 m (650–980 ft.) high. Dark-red spots on the image pinpoint oases fed by springs rising along faults.

The people of the Kharga oases battle sands that encroach on fertile areas (below). Old river channels funnel sand from the plateau to the north, forming parallel bands of barchan (crescent-shaped) dunes. Constant north winds drive the barchans southward as much as 10 m (33 ft.) a year, and attempts to stop or redirect them are rarely successful.

The perspective from space, a valuable aid for working on environmental problems, shows where the sand is coming from and where it is headed.

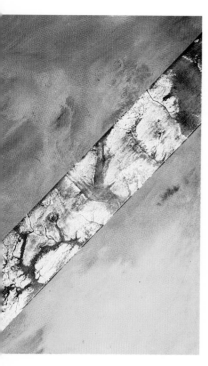

Tifernine Dunes

Dunes as high as 200 m (660 ft.) tower above surrounding gravel and rock in the Tifernine Dune Field, which is part of the Grand Erg Oriental, the eastern sand desert of Algeria.

The winds over the southern part of the Grand Erg are generally from the east, but seasonal variations can create complex shapes called star dunes at Tifernine. Migration of the sands to the south and west is blocked by the jaw-like Tassili-N-Ajjer, a cuesta or ridge, that tilts upward toward the south.

The windswept bedrock cuestas in this region form another type of desert, a rocky plateau known as a hammada (dark area at lower left). Lines on the hammada are ancient drainage networks, etched in the exposed rock by pre-historic rivers. These dry channels, along with others found throughout the Sahara, indicate that North Africa once had a humid climate.

Few people live in this region today because it lacks oases and has an extremely arid climate. But prehistoric rock paintings in the Tassili-N-Ajjer depict scenes in a moderate climate. The paintings span six thousand years and are a record of climate change. Earlier paintings depict hunters stalking game and herders tending cattle. Giraffes, wild sheep, hippos, and antelope are present. Later paintings show camels, suggesting arid conditions.

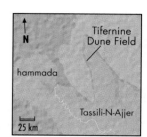

Vanished Rivers

Arid regions throughout North Africa display widespread patterns of dried-up river valleys called wadis that show how water flowed here in ancient times (above).

Scientists with the U.S. Geological Survey have used radar imagery collected by the Space Shuttle to study wadis in the eastern Sahara Desert. Because the desert sands are so dry (rain has not fallen for decades in some areas), the radar signal can penetrate the sand sheet and delineate structures below. In the illustration above, a shuttle radar strip is superimposed on a relatively featureless Landsat view of the northwest Sudan. The radar more clearly reveals ancient river courses hidden by the sands. Archaeologists use this information to locate sites of ancient human habitation likely to be found where water was once abundant.

A Visible Border

At the Berlin Conference of 1884-85, the European colonial powers decided to use physical features, such as rivers and mountain ranges, wherever possible to set bor-

ders between their overseas territories. Parallels of latitude or lines of longitude became borders when there were no distinctive physical features or when there was little knowledge of the terrain.

Ignorance and insensitivity thus divided peoples and disrupted traditional lines of migration and communication. One such straight-line boundary crosses this Space Shuttle photograph of the Angola-

Namibia border. A bilateral agreement between Portugal and Germany in the 19th Century established the boundary between Portuguese Angola and what was then German South-West Africa, (since 1990 the independent nation of Namibia). The boundary extended along 17°23'23.73" south latitude, from Ruacana Falls to the Cubango River and divided the traditional lands of the Ovambo people.

The Republic of South Africa, which administered South-West Africa after Germany's defeat in World War I, set aside the area south of the Angola-Namibia border as one of the reserves or "homelands" established for black Africans on lands of marginal value. Most of the estimated 1 million Ovambos lived there.

When this photograph was taken in August 1985, the boundary could be seen

because a fence was erected along the border to separate Angolan and Namibian Ovambos. The lighter color on the Namibian side of the border fence suggests more intensive grazing and firewood collection.

The dry lakebed of Etoshapan and surrounding smaller salt pans (lower right) appear white because of the presence of salt and other evaporites.

Namib Desert

Coastal deserts, such as the Namib of southwestern Africa, typically lie close to cold offshore waters that create a temperature inversion. In an inversion, a layer of warm air lies over a layer of cold air, limiting vertical air movement and thus inhibiting cloud formation. This atmospheric condition produces the intense dryness—technically, hyperaridity—found in the Namib Desert, in the Atacama Desert on the west coast of South America, and in similar deserts in northwest Africa, Baja California, and the west coast of Australia.

The Benguela Current flowing north along southern Africa's west coast carries cold surface water from Antarctica's Circumpolar Current. When humid air passes over the cold Benguela Current, condensation occurs and fog forms. The fog moves onshore, bringing the only moisture that the desert may receive for decades.

The rainless climate of the Namib Desert has led to specialized adaptations by plants and animals that live in the dunes. For example, one species of beetle (above) stands on its head to drink water captured from the coastal fog. Water condenses on the beetle's back and flows down into its mouth, allowing the beetle to drink from the fog borne by the morning breeze.

This Space Shuttle photograph (left) shows the Namib Sand Sea, the driest part of the Namib Desert, which extends for 2000 km (3,225 mi.) along the coast and contains several smaller dune fields. A number of dark circles of igneous rock are visible north of the Namib Sand Sea. The circle at the top of the image is the Brandberg Intrusion, a granite dome that was exposed by erosion of the Earth's surface more than 100 million years ago.

(following pages)

Sossus Vlei

The Tsauchab River intermittently flows down from the Naukluft Mountains of western Namibia and empties into the eastern Namib Sand Sea at Sossus Vlei. The occasional intrusion of the Tsauchab River has disrupted the linear pattern of dunes in the sand sea and formed a vlei, or marsh, separating the dunes. The star dunes on either side of the vlei range in height from 100 to 150 m (330 to 490 ft.). The dunes along the eastern margin of the Namib Sand Sea are a yellow-red color, and the deposits from the Tsauchab River appear blue and gray in this Space Shuttle photograph.

The Tsauchab drains a portion of the plateau of southern Africa composed of black and dark gray sedimentary rocks, such as shale and dolomite. The river transports gravel, silts, and clays down from the plateau, which give the riverbed and its small alluvial fan a light-blue color. The cause of the red color in some desert sands is still debated. One argument is that red sand is the result of stability and aging of sand dunes, while another argues that the red color is caused by higher humidity and higher temperature. Humidity and temperature in the Namib Sand Sea increase away from the coast, and are highest along the inland margin of the desert.

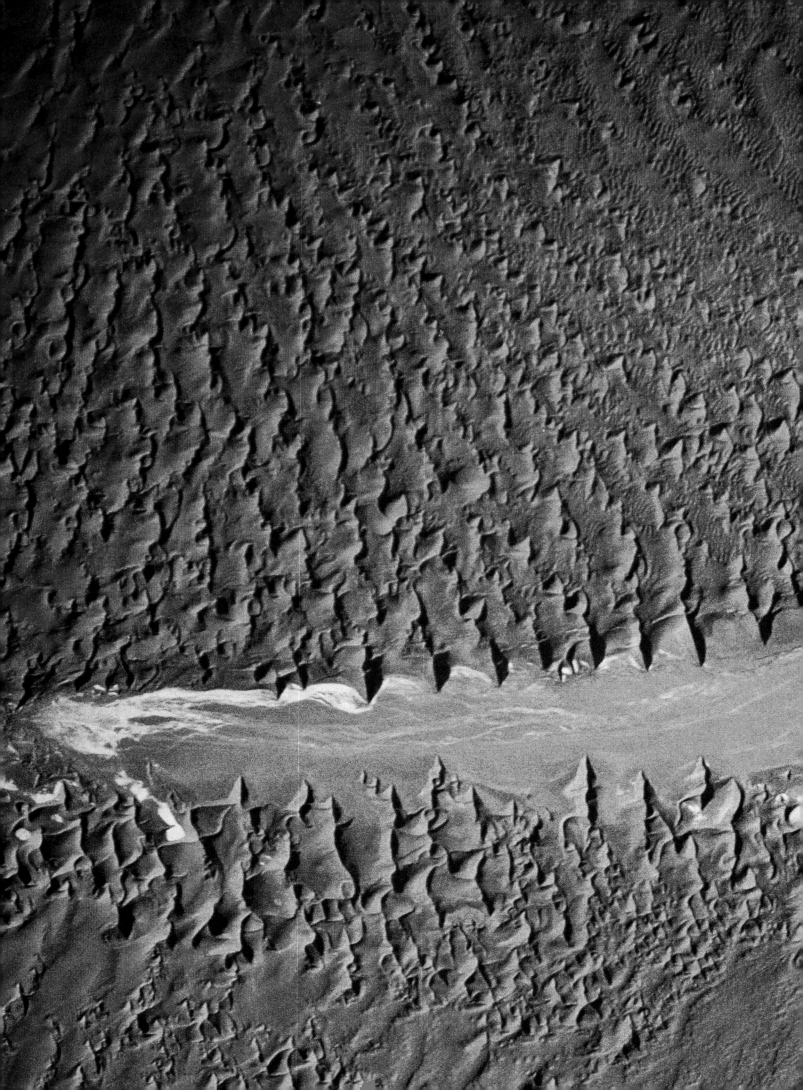

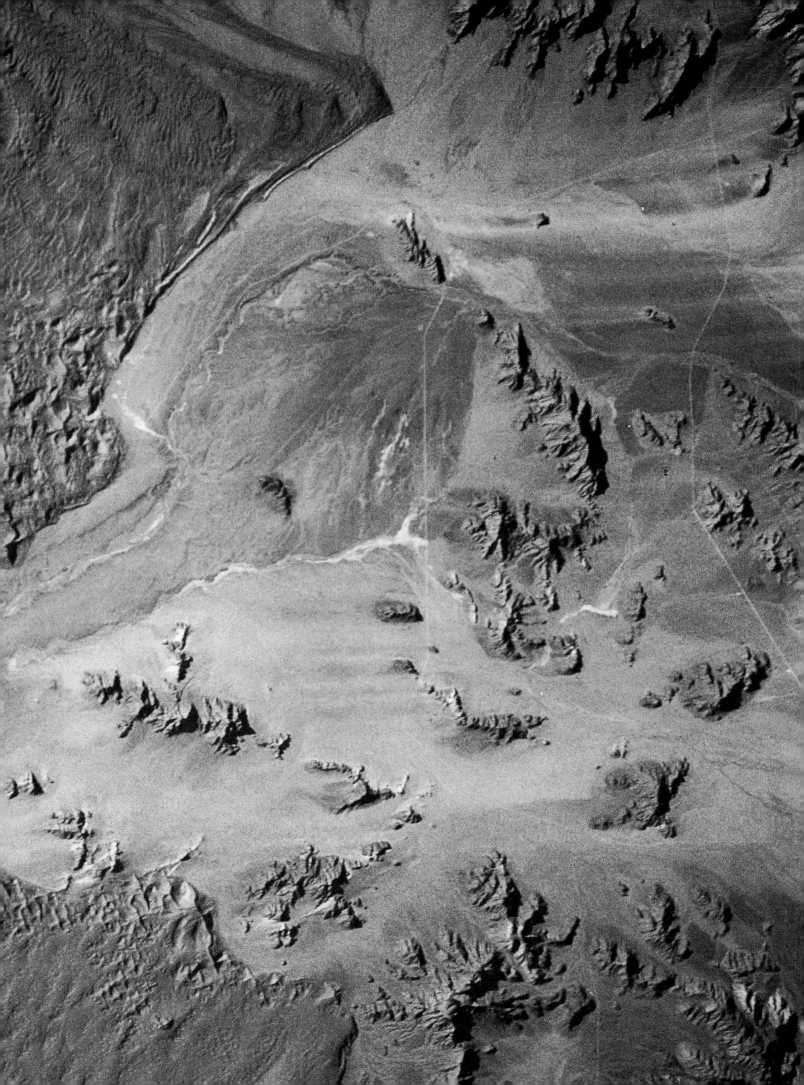

The Rivers that Wander

THE HUMID REGIONS OF AFRICA GIVE RISE TO SEVERAL OF THE world's major rivers, including the Nile, the Zaire (Congo), the Zambezi, and the Niger. Arid north Africa has only the Nile, which originates much farther south. South of the Sahara, several large river systems lace the continent. Some, such as the Nile and the Niger, are temporarily detained in large swamp-filled basins before they reach the sea. Others never reach the sea, flowing instead into interior basins and forming large, freshwater lakes, such as Lake Chad. In the more arid regions of southern Africa, rivers flowing into interior basins quickly evaporate, leaving broad salt pans on the basin floors. Occasionally, rivers end in swamps. The Okavango Delta is a wetland created as sediments were deposited in an internal drainage basin (see page 56). Though Africa is the Earth's second largest landmass, its rivers account for less than 10 percent of all the freshwater that flows into the planet's oceans.

The Nile, the world's longest river, empties into the Mediterranean Sea after its long journey across half of Africa. The river flows more than 7900 km (4,900 mi.) from its ultimate source in the mountains of Burundi, a small east African country, to its well-developed delta on the Egyptian coast. The Zambezi River drains south-central Africa and flows to the Indian Ocean. It crosses several significant natural and man-made obstacles on its route, including Victoria Falls and Kariba Dam. The latter provides electricity to the entire region. The Zaire (Congo) River is the largest African river to drain into the Atlantic Ocean (see page 57). Rapids and waterfalls along its course through Africa's rainy central interior provide much of the continent's hydroelectric power.

Throughout prehistoric and historic times small human settlements and large civilizations have been concentrated in Africa's river valleys. The earliest inhabitants of the Okavango Delta in the south, for example, were probably the Banoka or "River Bushmen," related to the San people of the surrounding Kalahari Basin. The Banoka traveled and fished the channels of the delta on reed rafts. The Hambukushu, who started emigrating from Zambia and Zaire about 1750, fled a series of oppressive conquests in their homelands. In the mid–18th Century the Bayei people took refuge in the delta from the strife-torn Lozi Empire. From their dugout canoes the Bayei still fish and hunt along the delta's channels.

About 99 percent of Egypt's population lives in either the Nile Delta or the Nile Valley. The pyramids and other ancient archaeological structures found along the Nile tell the story of past civilizations supported by the river's waters. Aswan, the site of the first of the six cataracts, or rapids, of the Nile, was the limit to upstream navigation on the lower Nile. The second cataract is now submerged beneath Lake Nasser.

Along Africa's river valleys people have found fertile soil for agriculture, fish, and game, a system of transportation, and, in more recent times, sites for dams and hydroelectric power stations. Power transmission lines and irrigation systems have extended the rivers' benefits beyond the floodplains, but the adverse effects these projects created have become apparent over the years.

The damming of the Nile at Aswan in 1971 has caused a depletion of fertile silts that were regularly deposited on the river's floodplain during the annual floods (see page 52). Irrigation projects tapping Nile tributaries have caused problems, such as soil salinization that occurs as the water evaporates in the fields and leaves salts concentrated in the soil. The damming of the Zambezi, which created Lake Kariba, flooded valleys that were home to more than 50,000 people and spread waterborne disease along the lake's shores. The unforseen impacts of past grand schemes for Africa's rivers—many sponsored and funded by industrial nations and international development agencies—clearly show the need for better long-term social and environmental assessments to balance the perceived short-term benefits.

Lake Nasser extends for more than 480 km (300 mi.) from the Aswan High Dam in southern Egypt southward into Sudan. The lake's shoreline is deeply embayed, the result of the impounded waters filling the eroded stream channels cut long ago in the surrounding plateaus.

The Nile

This mosaic of Landsat images shows the lower 1400 km (868 mi.) of the Nile River. The large red triangle is the Nile Delta, formed by silt deposits. The Greeks called this phenomenon a delta because its shape resembles the Greek letter Delta ($\triangle$).

The Nile River enters its delta and branches into two main distributaries, the Rosetta and the Damietta. Visible are Cairo, a blue-gray spot near the delta's apex; the Sinai Peninsula; and the Suez Canal, which connects the Mediterranean and the Gulf of Suez.

Upstream, a red band indicates crops cultivated on a floodplain, which narrows until it reaches Lake Nasser. The Western Desert (left) is sparsely populated by nomads. Farmers live in the desert's oases, which lie in depressions amid sand streaks and sandstone plateaus.

MEDITERRANEAN SEA
Nile Delta
Suez Canal
Cairo
Sinai Peninsula
RED SEA
Nile River
EGYPT
Aswan Dam
Lake Nasser
N
100 km

Aswan Dam

The Nubian Valley, a narrow gorge of the upper Nile River, is the site of the Aswan Dam. The original dam, completed in 1902, had a system of locks to make the upper Nile more accessible to navigation. In 1960 a more ambitious project was begun 6 km (nearly 4 mi.) upstream of the first dam. This involved construction of a dam 114 m (375 ft.) high. The Space Shuttle Challenger's Large Format Camera photographed the Aswan High Dam in 1984 (above). Roads, buildings, and the structure of the dam can be seen clearly.

Approximately 100,000 inhabitants of the valley were moved to New Nubia, a collection of government-built villages downstream of the dam. Very little arable land was submerged beneath the new lake because the steep-sided river gorge contained poor soils. However, many treasures of ancient Egypt were threatened by the rising lake waters, most notably the temple at Abu-Simbel, built about 1250 BC during the reign of Ramses II. In 1966, UNESCO funded the disassembly and transport of the temple and statues to a position 61 m (200 ft.) above the original site and beyond the new lake's shore.

The Aswan High Dam Project was completed in 1970, and Lake Nasser was named for the late president of Egypt. The project was built to provide a steady supply of irrigating water to the agricultural lands along the Nile, to provide electricity for all of Egypt, and to control flooding. Egypt can potentially expand its arable land by 810,000 hectares (2 million acres) by using water from Lake Nasser, and abundant electrical power is now available for industrial growth.

In the short term the project has provided benefits. But without the annual flood, new silts are not added to the floodplain and soil fertility decreases. Sediments previously deposited on the floodplain when the Nile spilled its banks are now accumulating behind the dam at Aswan, denying the rich silts to the floodplains downstream.

This Space Shuttle photograph (right) was taken in late September 1988 and shows the northern end of Lake Nasser when the water was high. Following the summer rains in the Upper Nile Basin and on the East African Plateau, the Nile rises to flood stage, which arrives at Aswan in July.

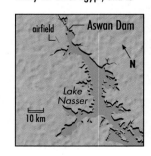

(following pages)

Senegal River

Rising in the highlands of west Africa, the Senegal River flows north into the Sahel. This SPOT image shows, in red, grasses growing in the Sahel during the summer monsoon. The blue-gray area along the Senegal's middle reaches consists of wet soil and backswamp on the river's floodplain. North of the river lies Mauritania and the southern margin of the Sahara.

The long sand streaks in Mauritania indicate the northeasterly wind direction. Near the end of the dry season, in spring, the Harmattan—a strong, hot, and dry wind—blows out of the Sahara bearing clouds of sand.

The time lag between the annual rains over the highlands to the south and the arrival of the floodwaters in the river's middle reaches makes it possible to have two harvests each year. During the summer monsoon the Fulani people of the region plant beyond the floodplain and rely on rainfall to water the crops. In the early autumn the Senegal floods, and when the waters recede crops of maize, sorghum, and rice are planted on the wet floodplain.

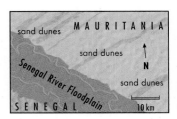

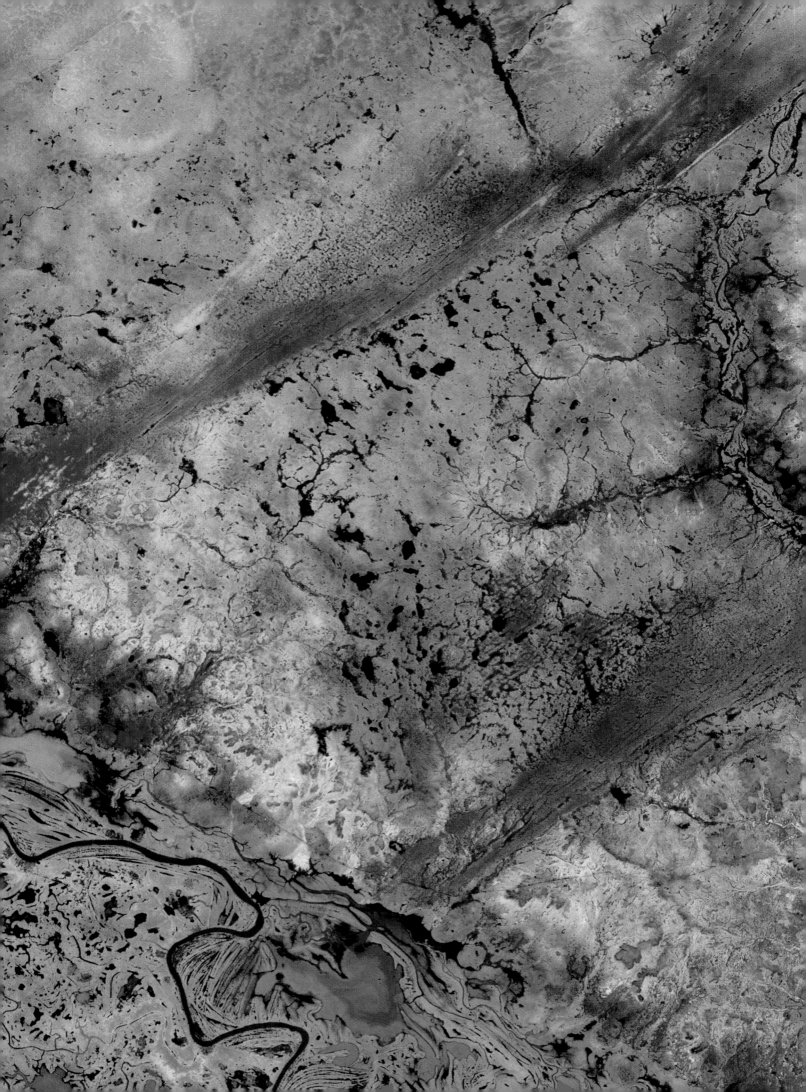

Okavango Delta

Like other major rivers in Africa, the Okavango flows into an internal drainage basin. But, unlike the other rivers, the Okavango does not continue to the sea.

The waters of the Okavango fan out when they reach a broad, shallow graben, or trench. The bird-foot shape is formed by channels diverging from the delta's panhandle, which lies in a narrow graben perpendicular to the larger trench. Sand dunes stabilized by vegetation form patterns of parallel lines surrounding portions of the delta. The main tributaries of the Okavango, the Cuito and the Cubango rivers, flow down from the central highlands of Angola into northwestern Botswana.

This Space Shuttle photograph was taken in early November 1985, about two months after the arrival of the annual flood at Shakawe, a village at the northern end of the panhandle. Dark patches are the perennial swamps where the dominant vegetation is papyrus and bulrush growing on dark beds of peat. Crocodiles, hippopotamuses, and lechwe, an African antelope species, live here.

Seasonal swamps, products of the annual floods, appear as lighter tones adjacent to the perennial swamps. Grasses dominate the delta's floodplain, the habitat of buffaloes, zebras, wildebeest, and elephants. On the periphery of the delta, in drier grasslands, live giraffes, springboks, ostriches, warthogs, hyenas, leopards, and lions.

The earliest inhabitants of the delta were probably the Banoka, who traveled and fished the channels on reed rafts and small dugout canoes.

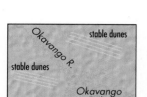

Okavango R.

stable dunes

stable dunes

Okavango Delta

20 km   N

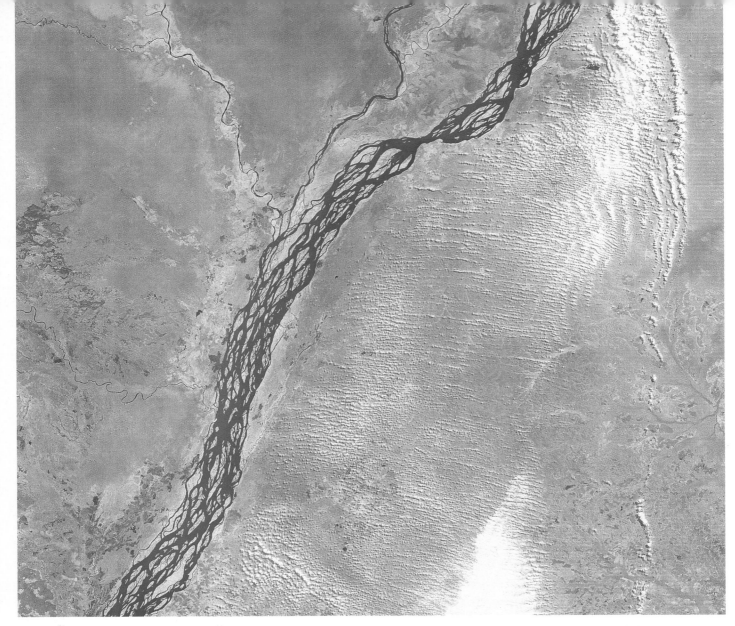

Zaire River

The Zaire (Congo) River rises in Zambia and courses northward in a great arc, finally flowing into the Atlantic Ocean on the coast of Zaire. The river runs for nearly 4800 km (3,000 mi.), mostly through thick rain forests. Year-round rains feed the river along much of its course.

Through some areas it serves as a vital artery for the transport of such minerals and products as copper, cotton, and coffee.

The upper Zaire, featuring lakes and waterfalls, stretches roughly 2700 km (1,700 mi.). The middle Zaire, sometimes nearly 13 km (8 mi.) wide, is the most navigable part of the river. Here it often separates into branches dotted with islands. This clear image (above), in a region of broad marshes, is a rare one; the river, flowing in a rainy climate, is generally under clouds.

The lower reach of the river begins about 400 km (250 mi.) southwest of this image, where two branches join to form the Malebo Pool. A series of rapids follows, and the river drops more than 245 m (800 ft.) over a distance of 350 km (220 mi.). A short stretch then flows to the ocean.

Crocodiles, turtles, and water snakes live in the river, and numerous species of fish are plentiful. A boy checks fish traps (right) at the base of the rapids of Boyoma Falls, upriver from the city of Kisangani in Zaire.

Madagascar

Topography, heavy rain, and intensive farming have combined to produce disastrous erosion on one of the largest islands of the world, Madagascar. The severity of the erosion shows in this image (opposite page) of the mouth of the Betsiboka River, where a large sediment plume (lower center) stains the blue waters of the Mozambique Channel, which separates the island from the African coast.

The sediment's red color is caused by the decomposition of gneiss, the rock which forms the island's central plateau, and from iron oxide and aluminum hydroxide in the soil. The braided channels (upper center) indicate that the river is overloaded with sediments.

Madagascar mimics its neighbor, Africa, by having a prominent scarp and plateau in the east and by gradually sloping to the west. The Betsiboka River and its main tributary, the Ikepa River, drain the central plateau. The drenching rains that fall on the island are produced by the southeast trade winds that bring moisture off the Indian Ocean. The trade winds are deflected upwards by Madagascar's eastern escarpment.

As the moist air rises and cools, condensation occurs, producing clouds and rain. The island's heaviest rains fall along the scarp and on the central plateau. Annual rainfall at the Bay of Antongil, located on Madagascar's northeastern coast, averages 3700 mm (146 in.). Cyclones occasionally lash the island's east coast with torrential rains and high winds.

The island was settled by seafarers who spoke Austronesian, a term that represents a group of Malayo-Polynesian languages. The Asian migrants crossed the Indian Ocean in outrigger sailboats and probably reached Madagascar between 1,800 and 2,500 years ago. Most of their descendants live along the east coast or on the central plateau, the areas receiving most of the island's rainfall.

Intensive slash-and-burn cultivation, overgrazing, and the logging of the tropical forest—all combined with a humid tropical climate—have resulted in the irreversible loss of fertile soils.

The central plateau of Madagascar was once covered with trees and grasses, though researchers are uncertain about the actual extent of the original tropical forest. Today much of the natural protective vegetation cover is gone. As a result, the tropical rains produce heavy run-off that deeply erodes the island's soils and washes them into rivers and streams (above).

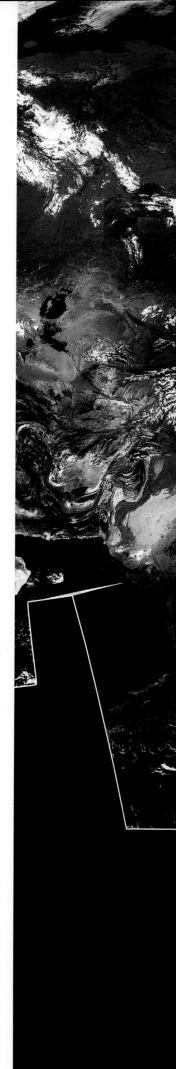

LARGEST OF ALL

ASIA, EXTENDING FROM THE URAL MOUNTAINS TO THE BERING STRAIT and from the Arctic Ocean to the equator, is the Earth's largest continent. The northernmost point of Asia's contiguous landmass, Mys (Cape) Chelyuskin, is only 1370 km (850 mi.) from the North Pole, while the southern tip of the Malay Peninsula lies 138 km (86 mi.) from the equator. Further extending Asia's range of latitude are islands that fringe the Asian landmass, such as Severnaya Zemlya in the Arctic Ocean and the islands of Indonesia, which lie well south of the equator. Nations of the Asian continent extend into Europe and face North America across a narrow strait. Similarly, the Middle East (see following chapter) occupies the southwestern corner of Asia and borders on Africa.

The range of Asia's climates and landscapes is also wide, and a description of the diversity gives a better sense of the continent's scale than conventional dimensions.. At Mys Chelyuskin the climate is polar: the land cover is treeless tundra, and the sun remains below the horizon for much of the winter. South of the Asian tundra lies the Siberian taiga, a snow forest of low conifers. Surprisingly, Siberia receives less winter snow than might be expected because the weather is dominated by descending dry air, known as the Siberian High (high atmospheric pressure).

Short grass prairie, or steppe, extends across the Asian continent south of the Siberian forests. The grasses of the steppes become sparse near Asia's great deserts, such as the Gobi in Mongolia and the deserts of Turkestan in Tadzhikistan (formerly part of the Soviet Union). Asia's mid-latitude deserts, unlike Africa's tropical and subtropical deserts, have bitterly cold winters. They remain dry because the high mountains of South Asia block the flow of moist maritime air from the Indian Ocean.

South of the mountains are several climatic regions and landscape types. The hot, subtropical Thar Desert, partially covered by scattered thorn trees and desert scrub, lies in eastern Pakistan and Rajasthan in northwestern India. In Southeast Asia tropical rain forests are found along the coasts. At higher elevations in Southeast Asia there is a distinct wet and dry season; here open-canopied monsoon forests dominate. Warm equatorial waters surround the large islands of Indonesia and the Philippines, where extensive tropical rain forests grow. Some of the larger mountainous islands also have grasslands on slopes or high plateaus. At New Guinea's highest point, Puncak Java (5030 m or 16,505 ft.), lies a small perennial ice field. It is about 440 km (275 mi.) south of the equator.

In the mosaic (opposite page) the low- and mid-latitude portions of Asia are more visible than the northern and equatorial regions, which clouds partially

NOAA AVHRR MOSAIC, EARTH
SATELLITE CORPORATION

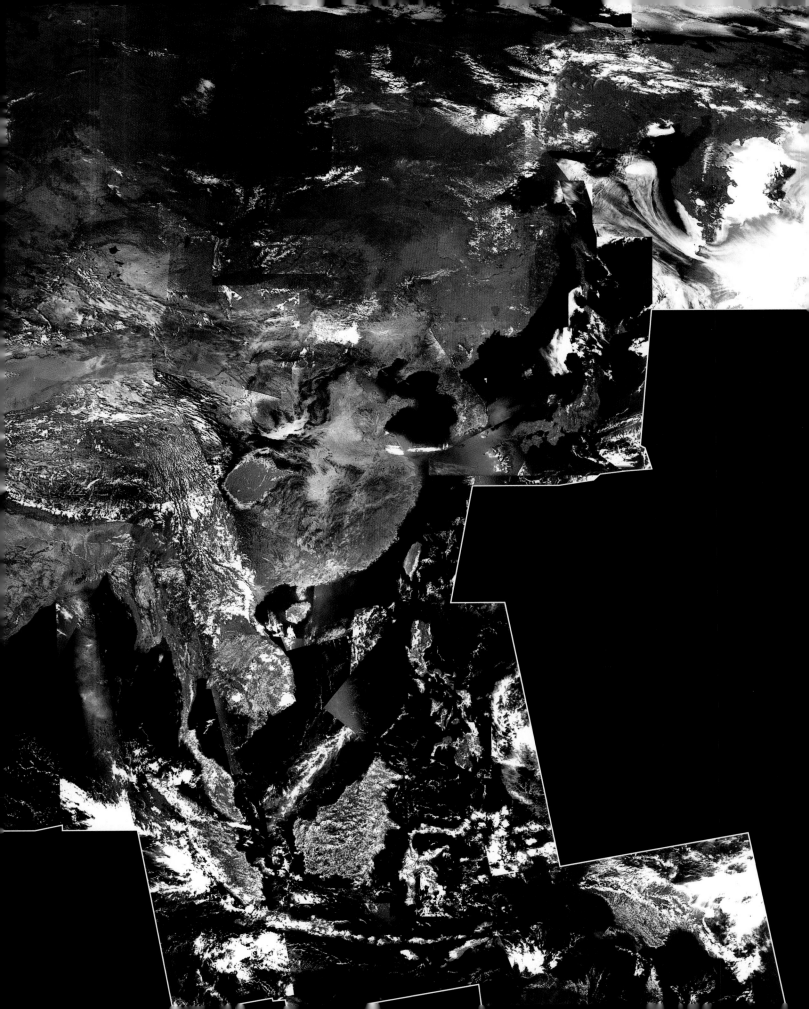

obscure. In Southeast Asia the rain forests appear bright red; in India the natural cover is less dense, giving the subcontinent a gray-pink tone. Sri Lanka is visible immediately south of India, almost connected to the subcontinent by the thread-like Adam's Bridge. The island is covered in patches of clouds that have risen over its hills and forests.

Two huge depressions called basins are prominently visible in China. The Tarim Basin of western China contains the Takla Makan Shamo ("sand desert"), another one of Asia's great deserts. The Sichuan Basin is a productive agricultural region in southwestern China, where the main crop, rice, is cultivated on terraces and on the basin floor. To the northeast, the Korean Peninsula projects towards Kyushu, the southernmost of Japan's four main islands. Japan is partially obscured by clouds, though Tokyo Bay is visible on the island of Honshu.

The line of the Himalayas' snow-capped peaks, separating India from Tibet, is one of the most distinctive features that Asia presents to a view from space. The Himalayas merge with the Hindu Kush, the Karakoram, and the Pamir Mountains to form the Pamir Knot, an isolated mountainous region that includes parts of Pakistan, Afghanistan, Tadzhikistan, and China. The Sulaiman and Kirthar ranges curve across western Pakistan, roughly paralleled by the Indus River and its floodplain. The mountains of south Asia were thrown upwards when, in one of the dynamics of the shifting Earth's crust, India collided with Asia. This is the best example of this type of mountain-building. (See page 65.)

Mountains can also have more explosive births, when volcanoes deposit molten rock and ash during eruptions. Volcanoes along the east coast of Asia and its adjacent island arcs form part of the "Ring of Fire," a chain of volcanoes that line the edge of the Pacific Basin. Many of the Earth's most dramatic, and deadly, eruptions occur along the Pacific Rim, such as the 1883 explosive eruption of Krakatoa, which obliterated an island and was heard more than 4800 km (2,900 mi.) away. Recent eruptions along the rim include Mount St. Helens in North America, Nevado Ruiz in South America, Mount Pinatubo in the Philippines, and the ongoing eruption of Sakura-jima in Japan.

The rivers of Asia are also well known for their dramatic behavior, flooding large areas and eroding great volumes of soil. Many of Asia's great rivers, such as the Indus, the Brahmaputra, the Chang, and the Huang, have their sources on the Tibet Plateau. The rivers of northern Asia rise in the mountains of Mongolia or eastern Siberia. Most of Asia's rivers flow into one of three ocean basins: the Indian, the Pacific, or the Arctic. Some rivers, however, find their way to inland seas, large lakes, or salt-encrusted marshes. When Asia's great rivers sweep over their banks, or when land shakes in an earthquake, the results are measured in the thousands of human lives lost.

More than half of the Earth's 5.3 billion people live in Asia, in the most densely populated nations in the world: Bangladesh has more than 8,400 people per 10 sq. km (3.9 sq. mi.), and the population of India is greater than that of the Western Hemisphere. Nearly one of every four people in the world lives in China. Absolute numbers inadequately describe the Asian population, however, because dense concentrations of diverse historic cultures exist while much of the huge landmass is sparsely inhabited.

ARCTIC OCEAN

BERING
SEA

RUSSIA

Yenisey River

Lena River

ASIA

SEA OF
OKHOTSK

p.72-73 Mount Zhupanova

MONGOLIA

p.84-86 Turfan Depression

p.81 Beijing

p.68 Japan

Amur R.

NORTH
KOREA

SEA OF
JAPAN

CHINA

p.75 Huang He Delta

Huang He

p.82 Hebei Province

SOUTH
KOREA

JAPAN

p. 88-89 Tokyo

p.67 Tibet Plateau

YELLOW
SEA

p.69-71 Mount Fuji

p.66 Mount Everest

p.76 Chang Jiang

Brahmaputra R.

NEPAL

BHUTAN

Chang Jiang

p.82-83 Shanghai

p.69 Mount Sakura-jima

BANGLADESH

p.76-77 Chang Jiang

Ganges R.

INDIA

p.78-79 The Ganges

BURMA

p.90 Hong Kong

TAIWAN

p.79 Flooding in Bangladesh

Salween R.

LAOS

Mekong River

BAY OF
BENGAL

THAILAND

PHILIPPINES

p.90-91 Bangkok

CAMBODIA

VIETNAM

SRI LANKA

SOUTH CHINA SEA

MALAYSIA

BRUNEI

PACIFIC OCEAN

SINGAPORE

INDONESIA

PAPUA
NEW
GUINEA

OCEAN

p.72 Lesser Sundas

Where the Earth Builds & Shakes

MOUNTAIN AND UPLAND TERRAIN COVER MUCH OF THE Asian continent. Asia's mountains display some of the most dramatic relief in the world, from the Himalayas, which soar to Earth's greatest heights, to the long, rugged chains of spectacular volcanoes on the Pacific Rim.

Asia's major ranges are young. The process that produced them was driven by the interaction of the Earth's plates that began about 130 million years ago, when a huge landmass consisting of several crustal plates lay in the Southern Hemisphere. This landmass, which geologists have named Gondwana, included today's Africa, South America, Australia, Antarctica, and India. The continents of the Southern Hemisphere moved to their present positions following the breakup of Gondwana, which was well under way by about 65 million years ago.

The Indian-Australian plate drifted north after separating from Gondwana. Part of the plate crossed the equator into the Northern Hemisphere, carrying a large piece of relatively light continental crust. The crust riding on the Indian-Australian plate crashed into the Eurasian plate and formed what we know today as the Indian subcontinent.

When India converged with Eurasia, the continental crust of each plate was crumpled and thrust upwards and downwards along the boundary of the collision. The convergence of plates produced the Great Himalayas, the highest mountain range on Earth. The Great Himalayas hold most of the world's peaks that are higher than 8000 m (25,000 ft.). Continental crust north of the Himalayas was uplifted about 5 km (3 mi.) by the collision, creating the Tibet Plateau.

The vast Himalayan belt of mountains is a classic illustration of what happens when two plates of continental material collide head-on. India's northward motion—slow (perhaps only a few centimeters per year) but relentless—resulted in the folding and upwarping of the Himalayan ranges in the early Cenozoic era, which began

A shallow, tropical sea floor formed the marine limestone of Mount Everest's summit 300 million years ago. Not until this century did anyone conquer the mountain to stand on its summit, the highest point on earth. Claiming that honor in 1953, at 8850 m (29,028 ft.) above sea level, were Sir Edmund Hillary, a New Zealander, and Tenzing Norgay, a Sherpa.

around 65 million years ago. (See opposite page.)

A different kind of plate interaction occurs on the eastern edge of Asia, where the Pacific plate and the Eurasian plate converge. The Pacific plate, carrying oceanic crust, is denser than the continental material of Asia. As a result, where the two plates collide, in areas known as subduction zones, the denser one plunges beneath the other.

Subduction zones are regions of disruptive earth movements and geologic instability, where deep ocean trenches often lie adjacent to arc-shaped groups of islands composed of chains of volcanoes. The volcanism is thought to be due to the melting of plate-borne material in a zone where the oceanic plate has reached great depths beneath the continental plate. Shallow and deep-seated earthquakes also plague these regions.

Island arcs concentrate along the western edge of the Pacific Ocean. One belt extends through the islands of Indonesia and eastward past New Guinea. Another system of arcs includes the Philippines, Taiwan, Japan, and the Kuril Islands, which lie on a line stretching between Japan and Russia's Kamchatka Peninsula.

The geology of these island arcs directly affects the lives of their inhabitants. Earth tremors and small quakes continually warn islanders that they walk on an ever-shifting land. Clouds of ash, repeatedly erupted from volcanic summits, threaten the health of those who breathe the polluted air. Catastrophic eruptions sometimes kill thousands: the huge 1883 eruption of the Krakatoa volcano in Indonesia produced a tidal wave that killed 36,000 people, and the disastrous 1976 earthquake on the island of Mindanao in the Philippines caused 8,000 deaths.

The plate movements that drive these events are very slow; although they can be measured, they seem unnoticeable. Yet to those who live where the Earth surges and shakes, the forces of terrestrial change can be sudden and powerful.

The Himalayas

Three-dimensional images can be manipulated to provide a re-creation of what happens when continents collide. These three perspectives were produced by combining land-cover data collected by satellite with elevation data from aircraft navigation charts, produced electronically in digital format. When merged, this data presents a view from different directions and different angles.

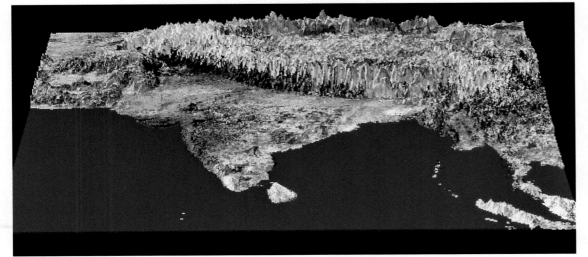

The first view (top) is from directly overhead, and gives very little information about the relief, or differences in elevation, within South Asia. In the other perspectives, from the south (middle) and the west (lower), the Himalayas rise abruptly and seem to wall off India from the rest of Asia. Elevation data is usually exaggerated in three-dimensional perspectives to give a better sense of an area's relief. Although the elevation data is exaggerated in these scenes, the images give a sense of the dramatic upheaval caused by a continental collision.

During World War II, the Himalayas, Sanskrit for "abode of snow," formed a barrier for Allied airmen attempting to supply China after the Burma Road was cut off. The airlift over the Himalayas, from Assay, India, to Kunming, China, was known as The Hump.

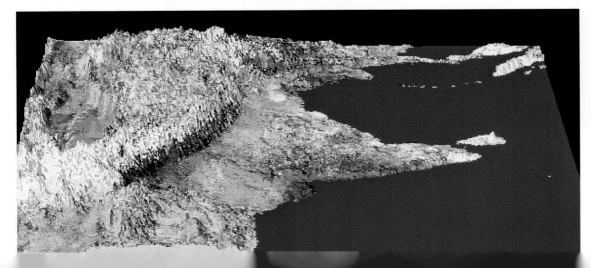

Mount Everest

The Tibetan name for Mount Everest is Chomolungma, Goddess Mother of the World. In 1852 the British Survey of India classified Mount Everest (8850 m or 29,028 ft.) as the world's highest mountain and—without the advice of the Tibetans—named the peak after the former Surveyor General of India, Sir George Everest.

Summer monsoons blowing north from the Indian Ocean bring heavy rains to the forested slopes of the Lesser Himalayas in Nepal. At higher elevations, heavy snows accumulate and avalanches frequently sweep down. The Himalayas block the northward flow of warm, humid monsoon air. Thus, north of the Himalayas, in the lee of the mountains, the Tibet Plateau remains dry and cold, with sparse grass cover. This color-infrared Metric Camera photograph, taken in December 1983 from the Space Shuttle, contrasts the rich vegetation (red) of Nepal and the lesser endowed land of Tibet (gray).

Mount Everest (left) straddles the border between the Kingdom of Nepal and Tibet, an autonomous region of China. During the late 19th Century Tibetan intolerance of foreigners prevented climbers from approaching Mount Everest, but beginning in 1920, climbing expeditions were permitted to enter Tibet from northeastern India.

These expeditions approached the mountain from the north, following the Rongbuk Glacier to a route up the Northeast Ridge. In 1924 two British climbers, George Mallory and Sandy Irvine, were last seen on this route. Climbers still debate whether they died before or after reaching the summit.

Sir Edmund Hillary of New Zealand and Tenzing Norgay of Nepal made the first successful ascent of Everest in 1953. They ascended via the Khumba Glacier and the South Col. The climbers had to approach from the south because, once again, Tibet had been closed to foreign visitors following China's 1949 invasion.

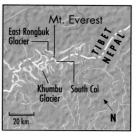

Mt. Everest
East Rongbuk Glacier
TIBET
NEPAL
Khumbu Glacier   South Col
20 km
N

Tibet Plateau

A remote highland of alpine tundra and grassland, the Tibet Plateau is surrounded by some of the Earth's great mountain ranges. The plateau's southern half is traversed from west to east by the Yarlung Zangbo Jiang (river). Known by its Indian name, the Brahmaputra, on the lower reaches, Yarlung Zangbo's source is the Chema-yungdung Glacier in southwestern Tibet.

Plateau elevation averages 4000 m (13,000 ft.), with a cold, dry climate, but the Yarlung Zangbo valley, with a more moderate climate, has a longer growing season compared to that of the higher northern part of the plateau. Cultivation of winter wheat, barley, and buckwheat is limited to river valleys, however, because of the lack of precipitation. Yaks provide transport, pull plows, and serve as the source of milk, meat, and leather.

Tibet is an autonomous region of the People's Republic of China. Its capital city, Lhasa, lies along the Lhasa He, a tributary river of the Yarlung Zangbo. Periodically throughout its history, the region has had political and cultural connections with China. Imperial China invaded Tibet in 1910. From 1911 to 1950 Tibet was independent, and the nomads of the plateau remained aloof to the war and revolution in neighboring China. In 1950 the People's Liberation Army of China invaded Tibet, forcing the region to join the People's Republic.

The spiritual and temporal leader of the Tibetans, the Dalai Lama, fled Tibet after an unsuccessful uprising in 1959, and resettled in India, where he was given sanctuary.

Since the 1959 uprising in Tibet, there has been civil unrest and ethnic fighting between Tibetans and Chinese migrants, many of whom have settled in Lhasa. The Chinese government has attempted to eradicate Lamaism, the Tibetan version of Buddhism, by destroying monasteries and persecuting Tibetan monks, or lamas.

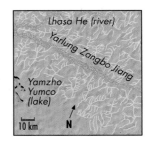

Japan

One hundred and seven individual Landsat frames, processed to produce an approximation of natural color, were combined to create this cloud-free view of Japan, which encompasses four major islands and thousands of smaller, often uninhabited ones. The large islands of Japan stretch from north to south over a distance equivalent to that from Maine to Florida, but are less than 450 km (270 mi.) wide.

Hokkaido, the least industrialized of the four major islands, preserves forests in its national parks. Sapporo appears as a brownish orange area near Hokkaido's southwest coast. Steep, rugged mountains run down the length of Honshu, Japan's largest island. Tokyo (center, right), Osaka (center), and the port of Nagoya on Ise Bay show up as deep orange. Hiroshima is located near the southern end of Honshu.

The other two principal islands, Kyushu and Shikoku, lie south of Honshu.

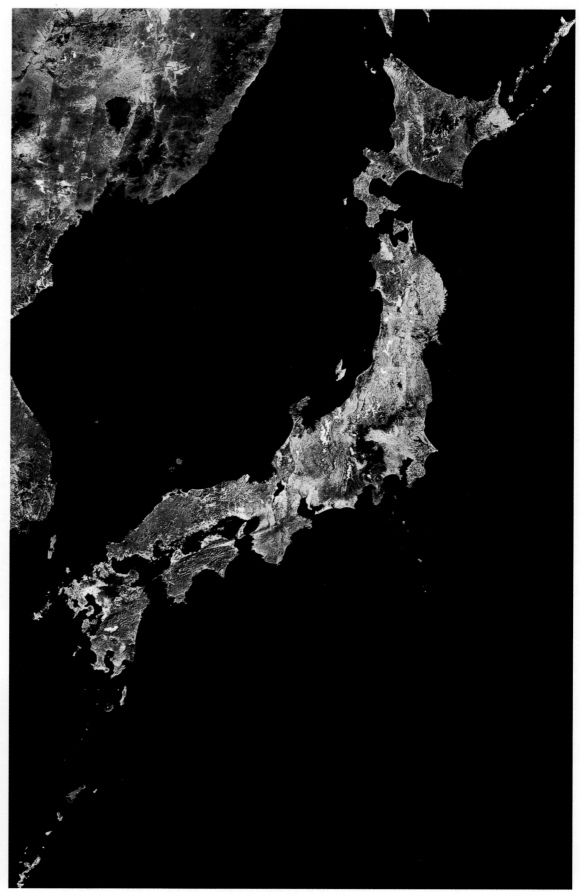

Mount Sakura-jima

Steam and ash are almost continuously emitted from Mount Sakura-jima, an active volcano in Kagoshima Bay on the southern edge of Japan. Ash sometimes blankets the nearby city of Kagoshima.

Said to have first formed in 708 AD, Sakura-jima has had a violent history of multiple eruptions and explosions through the centuries. It once stood as an island in the bay, but a massive eruption in 1914 filled in the 30- to 40-fathom channel that separated it from the mainland. Today a narrow strip of land joins the volcano to the coast. In a photo (right) an explosion from the 1924 eruption is seen.

In this SPOT image, the red color on the slopes of Sakura-jima indicates vegetation growing in the ultra-fertile volcanic soil. Fruits and vegetables (including 40-pound radishes) are cultivated in these regions. The unvegetated zones mark the sites of relatively recent flows, where plants have not yet had the chance to take hold.

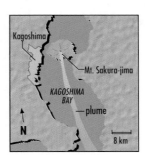

(following pages)

Mt. Fuji

Japan's highest mountain (reaching nearly 3780 m or 12,400 ft. above sea level), Mount Fuji rises about 90 km (55 mi.) southwest of Tokyo. It has long been renowned and admired for its beautiful symmetrical form.

An international tourist attraction, the snow-capped volcano has a summit caldera, or crater, of more than 600 m (2,000 ft.) across. It has not erupted since 1707. More than hundreds of thousands of people make the climb to the peak each year. These climbers usually take two days, timing their arrival at the summit so they can see the sun rising over the Pacific.

To produce a three-dimensional view, a Landsat image has been overlaid on topographic data. Populated areas are indicated by the light blue-gray color in the valleys and coastal lowlands dotted by lakes surrounding the volcano. This clear image is a rare one, for clouds usually obscure Fuji's summit.

Shinto, the ancient religion of Japan, teaches that kami, or divine existence, resides in aspects of nature that produce awe. Mountains, especially those as beautiful as Fuji, evoke kami. As Japan's tallest and most sublimely shaped mountain, Fuji suggests serenity to the Japanese. They revere it as a source and symbol of power.

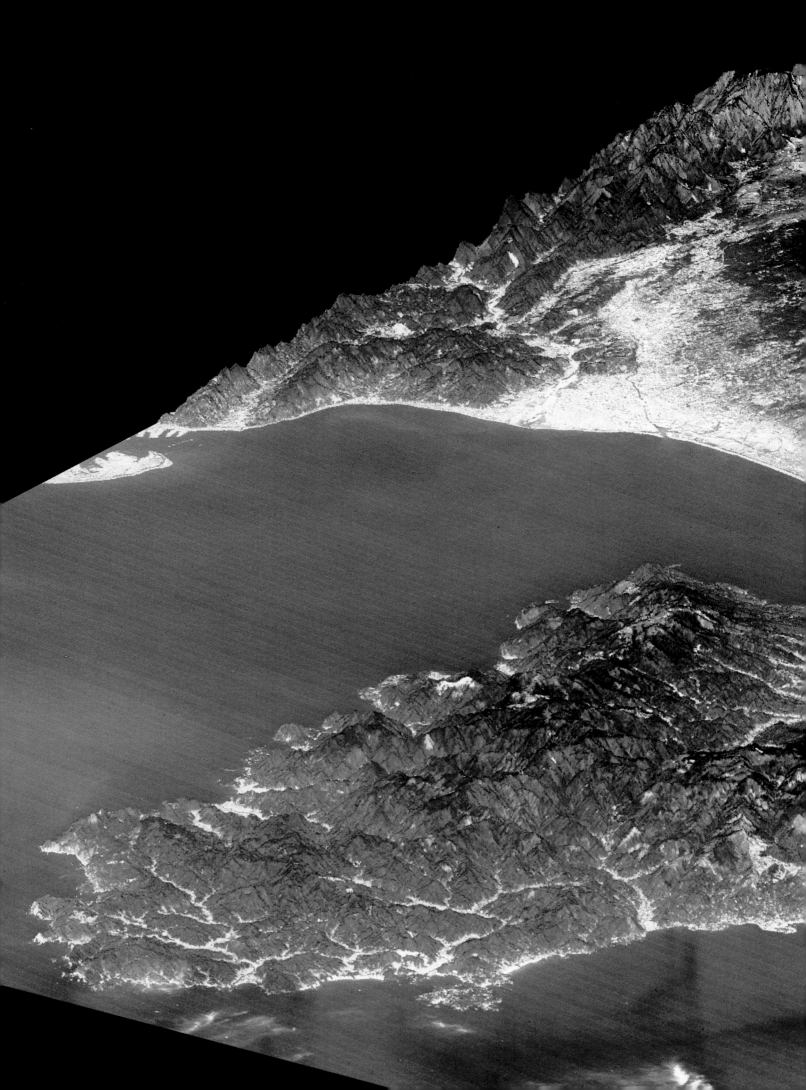

Lesser Sundas

East of Java stretch the Lesser Sunda Islands, some of the more than 13,000 islands of Indonesia that lie at a plate boundary on the edge of Asia's continental shelf. Volcanoes here are among the most active on Earth. Because the islands sit close to the equator, the climate is hot, rainy, and humid throughout the year.

Radar from the Space Shuttle reveals the terrain of several of the Lesser Sundas.

With the exception of Mount Api on the island of Sangeang, the volcanoes (below) are old and inactive. They are distinguished from the younger island by their rugged landscape, which flowing water has carved and eroded. Mount Api, which erupted as recently as 1988, exhibits flanks smoothed by young lava flows. The white streaks in the strait between Sumbawa and Komodo islands may be reflections of the sea, surface roughness churned by a mix of currents.

Mount Zhupanova

Lying on an extension of the island arc zones along the boundary of the Pacific plate, the Kamchatka Peninsula, at the far eastern edge of the former Soviet Union, has had a violent history of widespread volcanism and earthquakes. Remnants of more than one hundred volcanoes have been discovered in the area. Part of

the Pacific "ring of fire," most of them eroded and degraded by weathering and their own eruptions. Today about thirty complete volcanoes remain, and fewer than half of them are active. The Zhupanova volcano (opposite page), located near the southeast coast of the peninsula, last erupted in 1959. This elongate structure consists of several peaks, the highest of which reaches nearly 2930 m (9,610 ft.).

Large overlapping lava flows emanate from the smaller western peaks. Their ridged, ropy texture indicates viscosity generally found in eruptions of rocks rich in silica. Long natural levees border the flows. These structures are formed when lava rivers repeatedly overflow, building up congealed edges along their length.

The braided Zhupanova River, one of several rivers that dissect the volcanic plat-eaus, passes north of Mount Zhupanova and empties into a marshy peat bog along the coast. Lakes are common in the southern part of the peninsula, where heat from volcanic activity melts snow and ice and also forms hot springs and geysers. Lake Karymskoye (top, center) probably formed in the crater of an inactive volcano.

In 1975, Tolbachik volcano (this page, above left) on the Kamatchka Peninsula erupted in its most violent episode in history, spewing magma and gases steadily for three weeks. This extended eruption gave scientists plenty of time to study the volcano thoroughly.

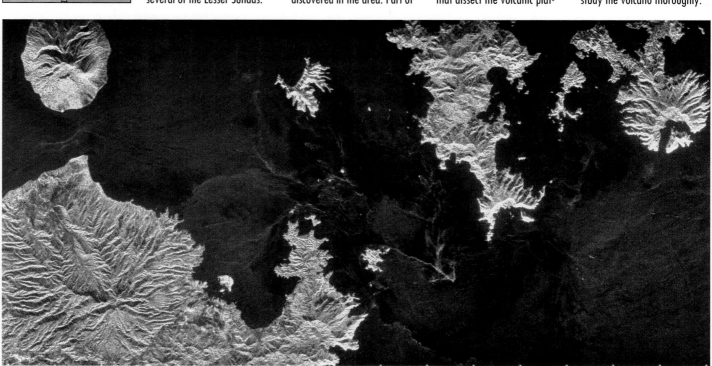

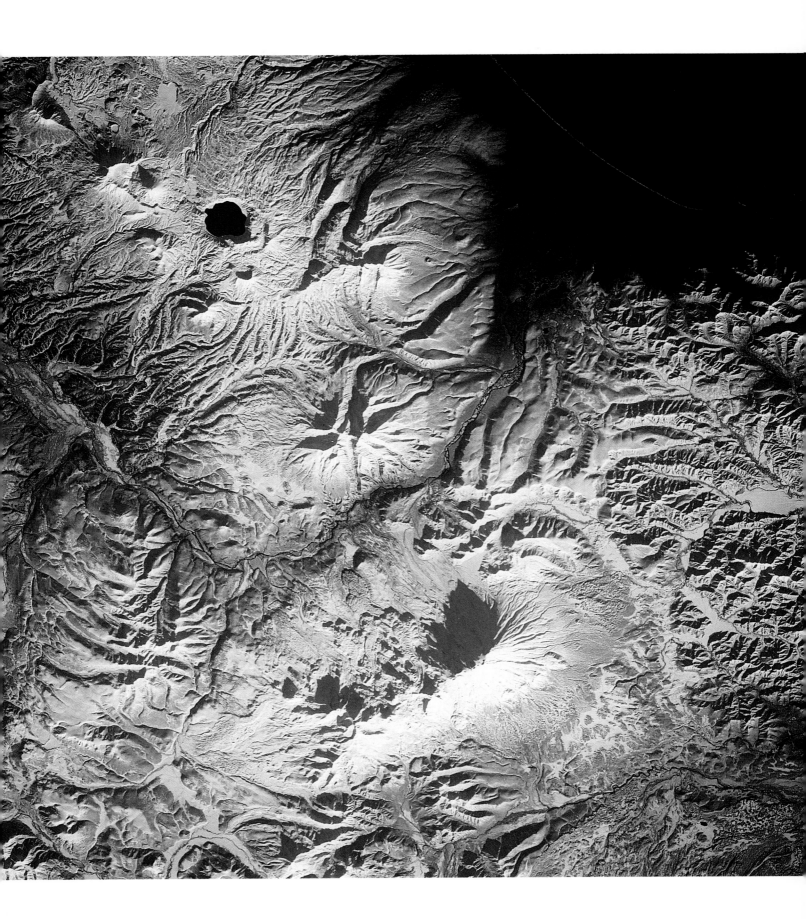

Flowing with Life & Peril

ASIA HAS A VAST NETWORK OF LONG AND POWERFUL RIVERS that have their sources in the central highlands. Northern rivers, such as the Ob and Lena, cross swampy forests and frozen tundra to empty into the Arctic Ocean. The lower reaches of these rivers may be frozen for as long as seven months of the year, and as the region's snows begin to melt, the rivers flood for long periods in the spring and summer.

In contrast, the rivers flowing through the dry areas of central and western Asia often have limited and intermittent flow. These rivers do not reach the oceans. Some, such as the Syr Dar'ya and the Ili, travel through internal drainage basins to empty into large inland bodies of water like the Aral Sea (see page 86) or Lake Balkhash. The waters of smaller rivers evaporate and fade away into desert sands.

Constant misuse of rivers and shortsighted economic planning doomed the Aral Sea, its fish, and its deltas during the past thirty years. In the 1930s what was then Soviet Central Asia became the site of huge cotton farms, and the two major rivers in the region, the Amu Dar'ya and the Syr Dar'ya, were tapped for water. Canals diverted water from the rivers to cotton farms hundreds of kilometers away.

The canals are not lined, and much of the water is lost to seepage into the porous desert soil. With almost all of the volume of these two rivers drawn off, very little water ever reaches the Aral Sea. By the late 1980s, the area of the sea had shrunk by about 40 percent.

The Huang He (Yellow), Chang Jiang (Yangtze), Mekong, and Red rivers flow to the Pacific Ocean. Lowland plains and deltas near these rivers are centers of agriculture and sites of some of the world's densest population concentrations.

The Huang He (opposite page) is known to the Chinese as "China's Sorrow" because of the frequent catastrophic floods along its lower reaches. Unlike the Chang Jiang to the south, the Huang He does not have a system of lakes along its course to buffer against floodwaters. When the Huang He floods, damage is caused not only from floodwaters but also from deposits of silt so thick that they can bury villages. But the silt also has great value: deposited on the floodplain, it renews the arable north China plain with nutrients.

The Huang He and the Chang Jiang (see pages 75 and 76) are among the world's longest rivers, and both have a millennial history of severe flooding. More than fifty major floods on the Chang Jiang have been recorded. In 1931, for example, long-lived heavy rains caused catastrophic flooding, destroying dams and displacing tens of millions of people.

Similarly, flooding is a tragic fact of life along the Ganges (see pages 78 and 79) and Indus rivers, which drain into the Indian Ocean. Where these rivers flow, millions of people live. They dwell along the shores and in the extensive deltas of the rivers, which are fed by glacial meltwaters and monsoon rains. Although the Ganges and Indus provide irrigation for intensive agriculture, heavy rains cause them to rise in disastrous floods, inflicting enormous damage and taking countless lives. Though the rivers have long brought death and destruction, they have also brought hope and life. India's earliest farmers are believed to have settled in the hills and plains along the Indus Valley in the sixth millennium BC. Rivers like the Indus gave the people life-sustaining waters, and in return rivers became sacred. The most revered of all the river goddesses was Ganga, namesake of the Ganges.

The Yangtze River (above) cuts through Tiger Leap Gorge, whose walls, according to legend, are so narrow that tigers could easily bound from one bank to another.

Asia's rivers provide transport, irrigation and fertile soils for vital food and commercial crops, centers for trade and industry, and sites for religious and cultural ceremonies. Yet they also bring destruction and tragedy to those crowded along their banks. Perhaps nowhere more so than in Asia are the lives and fates of the people so intimately intertwined with the course of a continent's mighty rivers.

Huang He Delta

The world's most active delta, growing at a rate of about 19 km (12 mi.) per century, builds at the mouth of the Huang He (Yellow River). The large blue and black rectangles are salt pans used for evaporating sea water. Cultivated fields, red in this false-color image, are visible further inland. The Huang He, the muddiest river on Earth, is China's second-longest river, running 5475 km (3,395 mi.) from eastern Tibet to the Bohai Sea.

The Huang He's yellow color is caused by the tremendous amount of yellow mica, quartz, and feldspar sediment, averaging 1.5 billion tons per year, transported by the river. Most of this sediment originates in the loess (wind-blown soil) plateau of north-central China carried down from the deserts of Mongolia and Xinjiang by northwest winds. Downstream of the plateau, the Huang He riverbed is higher than the surrounding floodplain, which has been built up by river-deposited silts.

BOHAI SEA

Huang He

N
salt pans

5 km   salt pans

Chang Jiang

Asia's longest river, the Chang Jiang (Yangtze), runs 5525 km (3,434 mi.) from its source high on the northern Tibet Plateau to the East China Sea. After flowing eastward across the plateau the river cuts a deep gorge and runs (right in image below) parallel to the Mekong (center) and the Salween (lower), two of Southeast Asia's major rivers. The Chang Jiang then turns northeast toward the Sichuan Basin, after having dropped 5183 m (17,000 ft.) in 2580 km (1,600 mi.) or about 3 m per km (10 ft. per mi.).

The Chang Jiang departs the Sichuan Basin and enters the Three Gorges. For 202 km (125 mi.) the river flows through limestone gorges with steep sides rising 400 to 600 m (1,300 to 2,000 ft.). The city of Yichang sits on the north bank of the Chang Jiang immediately downstream of the gorges (upper left, opposite page).

This Landsat image was made before the completion of the Gezhouba Dam at Yichang in 1988. Water contained behind the dam has submerged the rapids in the Xiling Gorge.

Plans exist for a much larger dam that will submerge the gorges entirely. The benefits of such a huge project are questionable when the costs are considered: loss of homes and cultural resources, and a host of environmental problems.

Below Yichang the Chang Jiang enters its lower reaches, crossing the East China Plain. As the river slows after passing through the narrow Three Gorges, it begins to deposit sediments, raising the height of the river bed and making the surrounding land vulnerable to flooding. Even though lakes act as buffers by absorbing the excess waters, the Chang Jiang produces a catastrophic flood about every fifty years.

The Chang Jiang basin, China's rice bowl, produces about half of China's agricultural output. One-third of China's population lives here.

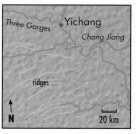

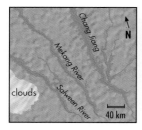

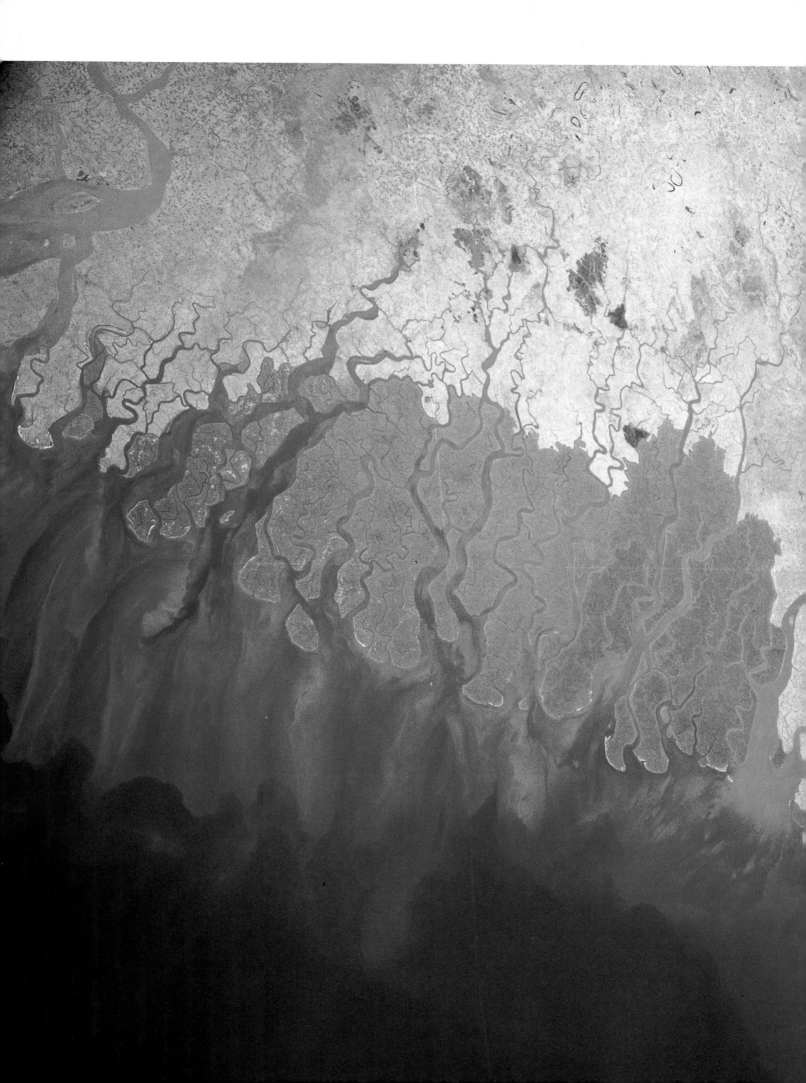

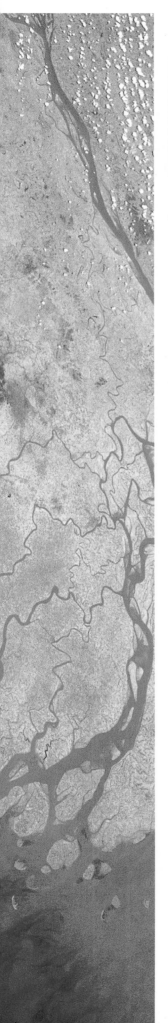

The Ganges

The Ganges River rises in the icy reaches of the Himalayas, the homeland of the Hindu gods. Myth says the god Brahma ordered the river goddess Ganga to descend to Earth from Mount Meru, abode of the gods. Infuriated, Ganga would have flooded Earth, but the Hindu god Shiva seized her waters in his hair, making her flow into seven weaker streams.

This most sacred river of the Hindus crosses through northern India and Bangladesh and empties into the Bay of Bengal. All along its course, from the Gangotri Glacier, a few miles from its source, to its humid delta, the Ganges is a focus of worship and ceremony for pilgrims who journey to the river. The mouths of the Ganges (infrared image at left) form a pattern of shifting intertwining channels, many clogged with silt that is deposited along the shores and into the Bay of Bengal. The red color, indicating vegetation, demarcates a region known as the Sundarbans ("beautiful forest"). This tract of brackish tidal waterways, reclaimed land, swamps, and forests is a refuge for wildlife, including deer and tigers, whose habitats

shrink drastically as farming intensifies. West of the Sundarbans is the Hooghly River, a Ganges distributary that flows past the city of Calcutta.

To the east Bangladesh lies in a humid lowland where the Ganges and Brahmaputra rivers converge to form a complex network of distributaries and deltaic deposits. Formerly known as East Pakistan, Bangladesh achieved independence in 1971. Separated from India and British rule in 1947, the nation was still joined politically with the culturally dissimilar West Pakistan. Dissatisfaction with rule from the west resulted in the 1971 uprising. With India's support, East Pakistan prevailed and became autonomous.

Annual floods have a devastating effect on densely populated Bangladesh. Dark areas in the upper image (right) indicate areas inundated by swollen rivers during the tragic flooding of September 1988. More than 2,000 people were killed, and 30 million were left homeless. In the lower image—a view of the region six weeks later —flood waters have receded. Another storm in January 1989 added to the worst flooding in a hundred years.

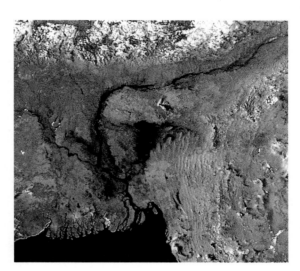

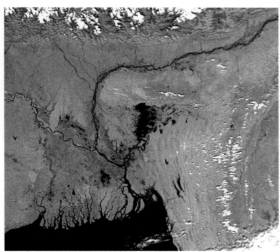

Humanity's Mark on the Land

FROM SPACE, HUMANKIND'S CONSIDERABLE IMPACT ON THE Earth is often not as obvious as the effects of the natural forces that shape the planet's surface. But spacecraft systems are continually enhancing our ability to see what people have done to the Earth. As smaller and smaller features come into view, the structures we build and their relation to the land around them become more evident. Since Asia is the most populous continent, humanity's presence there is well documented from orbit.

During the 20th Century, improved health practices in Asia lowered mortality rates and led to a rapid increase in population. Growth has slowed since the 1960s, but Asia's ever-expanding population and increasing urbanization could scarcely go unnoticed on space imagery. Closely packed clusters of towns and villages in India and China show up clearly, as does humanity's drain on the natural resources of these nations.

The crowded, sprawling urban centers of Asia often exhibit grid patterns of city streets, as well as parks and squares. (See pages 83 and 91.) Land reclamation projects are evident in crowded coastal cities, where growing industries and surging populations compete for limited space. Symmetrical patterns of irrigation systems and canals also indicate humanity's busy hands at work upon the land.

The intensity of modern farming often can be seen. Farmers of China's Hebei Province, for example, produce cotton, wheat, corn, and potatoes, and, with the exception of cotton, are able to harvest three crops every two years. This production has become possible with the use of chemical fertilizers, which has led to an increasing dependence on petroleum-based herbicides and pesticides. Such heavy use of the land is shown by the crowding of farmlands up against village boundaries. (See page 82.)

In China the emphasis has been on production and factory-building, with little thought for housing and urban amenities. The semiarid climate of the

Above, a cyclist passes by the Forbidden City in Beijing, the capital of Communist China.

Beijing region (see opposite page), combined with such industries as steel and chemical production, created a serious air-pollution problem. The region's water resources can barely meet the demands of industrial and residential users. Most of Beijing's residents still live in hutongs or compounds. Automobile, truck, and bus traffic, along with some 3 million bicyclists, jam Beijing's streets every day. But road improvement and construction has not kept pace with the increases in population and traffic.

Shanghai, home of a skilled labor force, is a metropolis of heavy industries, steel production, and petroleum refining. Sprawling factory complexes make the city one of the world's noisiest and most polluted urban areas on Earth. Factories occupy 25 percent of Shanghai's area. The lack of open green space within the city is evident in images from space. (See page 82.) Shanghai's present overcrowded and dirty conditions are in many ways similar to the large centers of the Industrial Revolution in Europe during the 19th Century.

The Chinese government has taken a balanced approach to rural and urban development, though the largest urban areas suffer from overcrowding and severe pollution. The government has dispersed industry to small cities in rural provinces. Migration to cities was even made illegal. Probably the harshest measure came after the Cultural Revolution (1965-68) when thousands of educated urban youths were forced out of China's cities to rural areas, often in remote western China, to work as laborers.

The effect of Asia's billions of people on the land surface is obviously substantial. Yet, comparing the scale of even the largest city to natural structures such as the island arcs, the span of the Himalayas, and the length of the great rivers, may help us to appreciate the power of the natural forces that are continually molding the face of the Earth, and to put human actions into a natural perspective. With respect to the planet, humankind's infrastructure is small—with large repercussions.

Beijing

The capital city of China and an important industrial center, Beijing has a population of about 10 million people. The blue-gray mass at the center of this SPOT image is urban Beijing. Surrounding suburban and rural areas appear red-brown because of greater vegetation cover. Urban areas show up dark on some false-color satellite images because many road and building materials, such as asphalt, are not very reflective in visible and infrared bands. The city was first given the role of capital in 1153 during the Jin dynasty. In 1949, after the Communists took power, Beijing became the capital of Communist China.

The Forbidden City, which contains the old Imperial Palace and the Gate of Heavenly Peace, is outlined by a rectangular moat near the center of Beijing. Immediately to the south lies Tiananmen Square, in June 1989 the site of a clash between pro-democracy demonstrators and the army.

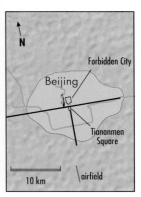

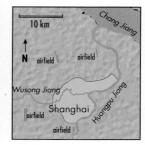

A Myriad of Villages

Closely packed villages in China's eastern Hebei Province glitter in this Space Shuttle radar image (below). The distribution of villages is about one per square kilometer. The brightness of so many small villages is caused by sheet-metal roofs, which reflect radar pulses transmitted from the Shuttle. The edges of cultivated fields appear between the villages. No space is left to waste. The darker area along the river is a onetime lakebed now being farmed.

The clusters of bright areas are the larger towns and cities. Dezhou, a city on the Grand Canal, is the most obvious of these bright areas. The towns' workers are employed in light industries, such as textiles and food processing.

The Grand Canal, the world's longest constructed waterway, is outlined by two roughly parallel lines as it approaches Dezhou from the south. The white lines are created by the reflection of the radar signal off the levees lining the canal. Pictured above, the Grand Canal cuts through the town of Wuxi, where it is used as a main street.

Shanghai

The Huangpu Jiang, Shanghai's arterial river, snakes along the eastern edge of China's most populous city.

Located near the mouth of the Chang Jiang (upper right in image on opposite page) on the East China Sea coast, Shanghai originated as a collection of fishing villages on the Huangpu.

When a port was established at the confluence of the Huangpu and Wusong rivers in 1074 AD, during the Song Dynasty, government offices were opened in the port to regulate the flow of goods along the coastal waterways. The fishing villages and the port were eventually consolidated into the zhen, or municipality, that became Shanghai, today the country's most productive industrial center, busiest port, and home to about 12 million people.

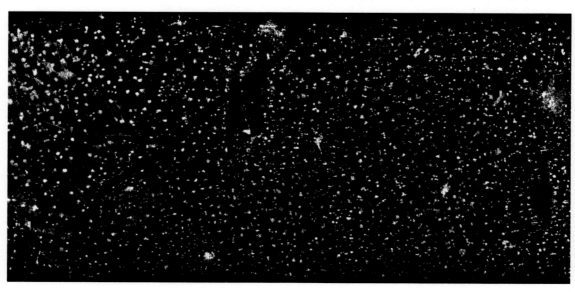

(previous pages)

Turfan Depression

The snow-capped Heavenly Mountains (Tien Shan) dominate this image of the Turfan Depression. The Tien Shan extend from the Pamir Mountains in Tadzhikistan (formerly part of the Soviet Union) to north-central China. A ribbon of red marks pine-forested northern slopes, watered by summer rains and melting snow. The arid southern slopes are grass-covered and have few stands of trees, most of them confined to deep, shady valleys.

The streams that drain the southern slopes flow onto a gravel plain, or gobi. In years of heavy snow, temporary streams flow across the gobi into the Turfan Depression. The dry channels of these streams can be seen on the dark gravel of the gobi. The faint line of a railroad disrupts the pattern of stream channels (right).

Water that seeps into the porous gravel recharges the groundwater, which flows beneath the gravel, supplying springs along the lower margin of the gobi. Cultivated land (red) fringes this area, making Turfan an agricultural oasis near the edge of one of the Earth's great temperate deserts, the Takla Makan. To the east of the oasis is a sand desert, or shamo, comprised of large, complex dunes.

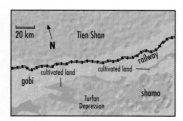

The Aral Sea

The Aral Sea lies in a drainage basin between two former Soviet republics, now the independent states of Uzbekistan and Kazakhstan. The remote desert region is far from the polluted industrial cities of eastern Europe. In spite of its isolation and lack of heavy industry, however, the Aral Sea has become the site of one of Earth's greatest environmental disasters.

The two major rivers that enter the Aral Sea have been tapped for irrigation. In this scene from the Soviet Resource-01 satellite (opposite page), damage is clearly visible where the irrigated fields and salt wastes collect in the two river deltas. A narrow ridge of exposed sea floor now divides the Aral Sea into two parts, northern and southern. Dried seabed is also exposed along the eastern and southern coasts, where white salts have been deposited as the Aral Sea evaporated. These salt wastelands are a source of acidic rains and windblown salt dust that plague surrounding oases by damaging buildings, farm equipment, and crops. Clouds of salt dust choke and poison people in the area, while excessive use of toxic herbicides and pesticides poison the soil and nearby small lakes. Particularly toxic are the waters draining from the region's extensive cotton fields.

Increased salinity has killed the sea's fish and shut down the region's formerly productive fishery. The boats of thousands of jobless fishermen lie abandoned on the seabed, scattered like toys in a sandbox (above). Frozen fish are now shipped in from the Pacific fishery for processing at plants that had once handled the catch of the Aral Sea.

The emergence of independent states in the former Soviet Union may end environmentally insensitive development. At present, United Nations experts are looking at ways to stabilize the Aral Sea and perhaps undo some of the damage. But the new republics, hard-pressed for foreign currency, may have to sacrifice their cotton exports to save the Aral Sea.

The second-largest body of water in Asia, the Aral Sea is rapidly becoming a sea of sand and salt. Since 1960, more than 40 percent of its surface area has been lost, leaving water that is too saline for the marine life it once supported.

Tokyo

The intensive development of central Tokyo is portrayed in this satellite image (opposite page) from the Japanese MOS-1 satellite. Wharves line Tokyo Bay. Few parks or grassy lands (red/orange colors) appear, except for the gardens (at left, outlined by moats) around the Imperial Palace, home of the Emperor. The present palace was built in 1968 to replace the 19th–Century structure built on the site of the old Edo Castle and later destroyed by air raids in World War II.

Tokyo is Japan's commercial, economic, and industrial hub, a crowded, fast-paced city of concrete and skyscrapers, juxtaposed with older areas of small shops and traditional architecture. If combined with the contiguous port of Yokohama, the Tokyo metropolitan area is home to more than 11 million people.

At the beginning of the 17th Century, the shogun (military ruler) Tokugawa Ieyasu rebuilt a castle in the town of Edo (meaning estuary) for his own use, and from there ruled Japan. When the Tokugawa shoguns were deposed in the mid–19th Century, the imperial family set up rule at the castle and called the city Tokyo (literally meaning Eastern Capital).

The city was almost destroyed twice. An earthquake in 1923 devastated the region around Tokyo and killed more than 140,000 people. U.S. bombing raids in 1944 and 1945 wiped out half of the city. One fire-bombing raid alone killed more than 83,000 people. Rebuilt after the war, Tokyo has expanded rapidly, physically and economically. Above, Tokyo City Hall's twin towers rise with the skyscrapers of Shinjuku Ward.

The city marks the northern end of what is termed the Tokaido Megalopolis, an intensely developed urban population center stretching southward to Osaka. It is named after the Tokaido Road, which runs between Tokyo and Kyoto and has a long history as an important thoroughfare. As Japan became a manufacturing and industrial power, its population shifted in great numbers out of the rural areas to crowd the cities of the megalopolis. The spreading urbanization and dense concentration of population in Japan's narrow coastal plains has resulted in tremendous environmental stress on the air, water, and soils of the region.

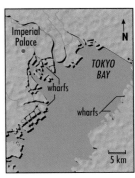

Hong Kong

A blue-gray mass near center left in the SPOT image below marks the Chinese city of Shenzhen, which lies along the border between China and the British Territory of Hong Kong. Hong Kong takes its name from an island that was ceded by imperial China to Great Britain in 1842. The island appears at the bottom of the image; Victoria, its capital city, fringes a narrow stretch of land overlooking Hong Kong Harbor. The hilly terrain of Hong Kong Island restricted building to a narrow band along the north shore.

The lack of available land, however, has not prevented Hong Kong from becoming a major force in Asian, if not global, banking and commerce. Land reclamation has provided a partial solution to the land-shortage problem. The single runway of Hong Kong's international airport is built almost entirely on reclaimed land.

Across the harbor, the city of Kowloon fills most of the level land below the hills of the Kowloon Peninsula. In 1898 Great Britain obtained a 99-year lease from China for land north of Kowloon, an area that became known as the New Territories. The expiration of that lease in 1997 will deprive urban Hong Kong of its neighboring dairy and vegetable farms and its water, which comes from reservoirs within the New Territories. Without the New Territories, Hong Kong is not viable as a foreign territory or an independent city-state on the edge of China. Although the British government has agreed to return Hong Kong Island and Kowloon to China, Hong Kong will have a special status, China says, that will protect its free-market system. Left, in January 1990, the city celebrates Lunar New Year.

Bangkok

A grid of rice fields and canals, or khlongs, etch the outskirts of Bangkok (opposite). Two deeper green parallel lines mark its airport. Bangkok is visible on the east side of the Chao Phraya River, upstream of the large meander. A smokestack plume, appearing as a sharp blue line, blows toward the center of Bangkok from an industrial area near the river. A cut-off meander, once a channel of the Mae Nam Bang Pakong River, is visible as a light band surrounding vegetation at lower right. Salt pans and shrimp ponds can be seen as green strips along the coast of the Gulf of Thailand.

Bangkok is the dominant urban area in Thailand with a population of about 5.5 million, almost two-thirds of the country's urban population. King Rama I founded the city in 1728 as an easily defendable site for his grand palace. The capital city is called Krung Thep (City of Angels) by most Thais. Rama I protected his palace by digging a khlong between bends in the river, thus forming a moat. Khlongs also formed Bangkok's first transportation network. But when automobiles arrived, many khlongs were filled in to create streets. Because the khlongs also drained the low-lying city—a function overlooked in the automobile age—flooding frequently occurs now after the afternoon rains.

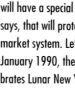

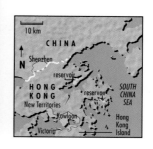

Bombay

Angular structures and networks of roads visible on the SPOT image attest to the extremely intense development of Bombay, one of the most densely populated cities of the world. The tallest buildings in India are found at the heavily commercialized Nariman Point, which juts out near the end of the peninsula. Originally a group of seven islands, Bombay was consolidated into a narrow peninsula by large-scale land reclamation projects in the 1860s. Along its east coast lies a picturesque and bustling harbor that has helped make Bombay a center of trade, manufacturing, and commerce.

The city has long been known for its textile and cotton industries. Bombay acts like a magnet, drawing large numbers of people from across India in search of jobs. From a population of only about 85,000 in 1900, Bombay grew rapidly to more than 1.5 million by the 1940s, and jumped from about 4 million in the early 1960s to more than 10 million by 1985. Bombay is India's most cosmopolitan city, a home to people of many countries and cultures. Though the city thrives culturally and commercially, severe overcrowding and intensive manufacturing have created rampant problems of homelessness, poverty, and pollution.

Rajasthan

Rajasthan (Land of the Princes) is a state in northwest India comprising the Thar Desert in the west and the forested Aravelli Mountains to the southeast. The desert and more fertile desert fringe areas of Rajasthan are among the most populated arid lands of the world. The population density is illustrated in this satellite image by the closely packed distribution of towns and villages (red spots). Light-colored halos around the towns and villages are areas where vegetation and scrub have been removed and soils have been damaged by overgrazing and intensive farming.

Prominent on the image is the shallow Sambhar salt lake. Its water level fluctuates from the seasonal rains draining down the Aravelli Mountains.

On the east end of the salt lake are the evaporation pools of the salt refinery. Salt works have been functional at Sambhar since the 16th Century.

The number of people and livestock inhabiting this area has been increasing since the early 1900s. The population growth has produced severe stress on the fragile ecosystem, especially in the fringe areas, where sparse natural vegetation helped to stabilize soils and dunes. Loss of vegetation has triggered dust storms, dune movement, and sand encroachment on fertile areas.

Scientists in Rajasthan use satellite imagery to observe changes in vegetation cover and sand distribution to plan ways to stop the land degradation. Among proposed strategies to diminish the stress is protection of limited forest lands near the mountains to reduce the environmental impact of the ever-growing population's need for wood.

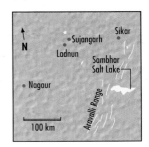

History's Pathway

THE TERM MIDDLE EAST WAS FIRST USED BY POLITICAL GEOGRAPHERS TO describe the region's location in relation to Europe, but just what is meant by the term Middle East has varied, producing many political and cultural definitions. Though frequently described as a crossroads, the Middle East has been far more than a mere intersection for invaders and traders. The area has been a source of fundamental developments in human history, including early agriculture, since the world's oldest civilizations were established by cultures that farmed the fertile floodplain soils of the Tigris and Euphrates rivers.

The arid and semiarid climate of the Middle East made irrigation essential to cultivation. As the scope and intricacy of the irrigation systems increased, complex social organization evolved. Towns were established as administrative, commercial, and cultural centers. Laws were codified and recorded.

In dry climates, where the amount of precipitation is low and the rate of evaporation high, irrigation can build up salt in the soil, which destroys soil structure and reduces crop yields. Agricultural output gradually declined in Mesopotamia as a result of salinization, as well as the repeated destruction of the irrigation systems by hordes of invaders. By the 12th Century large-scale agriculture had ceased in Mesopotamia, and the lowlands were being used as pasture by nomads.

Because much of the Middle East lacks vegetation cover and has an atmosphere that is generally cloud-free, the region yields fine views from space. The surface features and complex geology of the Middle East can be extensively observed and interpreted, as this 1990 mosaic reveals.

Geology has shaped the Middle East's landscape through rifting, uplifting, erosion, and deposition. Geologic processes, acting over millions of years, have also created the great pools of oil that lie beneath the desert sands and the shallow waters of the Persian Gulf. When this oil was tapped in the early 20th Century, the economy of the Middle East was changed more radically than at any other time since the demise of the great ancient civilizations.

Between Africa and Asia, close to the southeastern corner of Europe, lies the Arabian Peninsula, mostly covered by light-colored sand seas. Ad Dahnā' is a 1200-km (740-mi.) arc of sand that curves across the center of the peninsula. The arc originates in An Nafūd, a desert in northwestern Saudi Arabia, whose sands are transported down the peninsula by westerly and northwesterly winds. Directly south of An Nafūd, sand flow is obstructed by the Najd Plateau. The Najd, dropping off gradually to the northeast, eventually becomes low enough to allow winds to transport sand around its edge. Ad Dahnā' outlines the plateau and connects the Nafūd with the Rub al Khali, known as the Empty Quarter, a

COURTESY OF EROS DATA CENTER,

NATIONAL MAPPING DIVISION,

U.S. GEOLOGICAL SURVEY,

DATA COURTESY OF NOAA

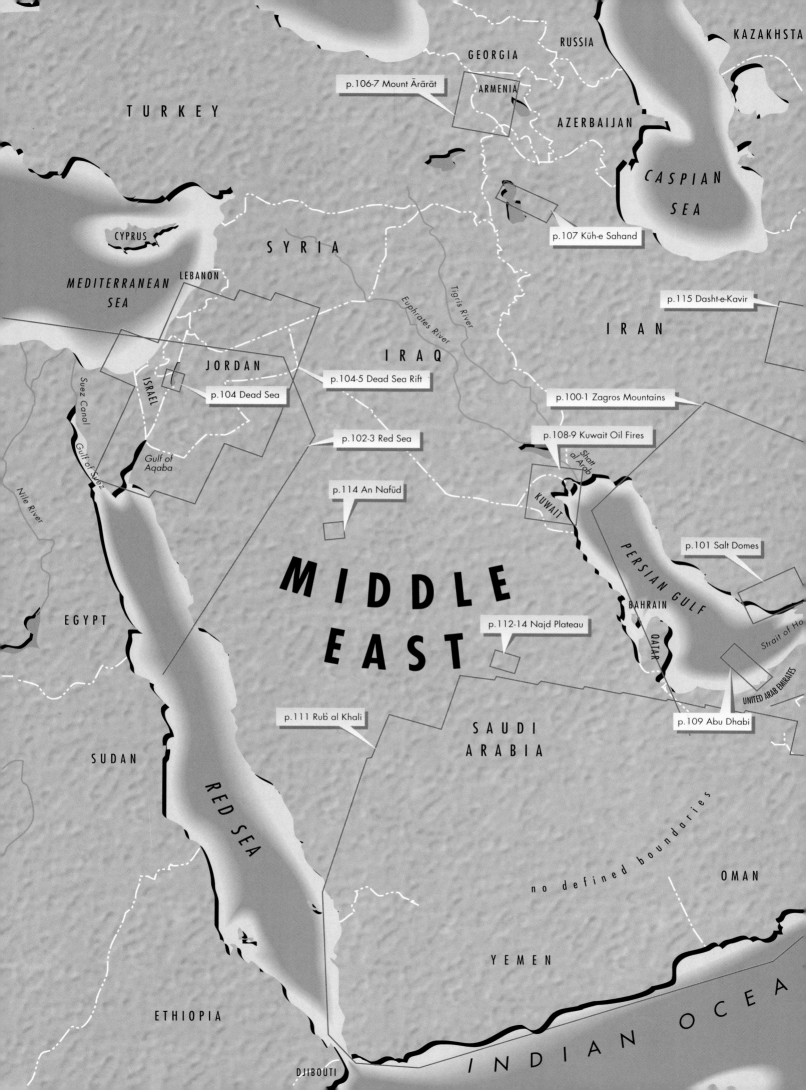

TURKEY

GEORGIA

RUSSIA

KAZAKHSTA

p.106-7 Mount Ārārāt

ARMENIA

AZERBAIJAN

CASPIAN SEA

CYPRUS

SYRIA

p.107 Kūh-e Sahand

MEDITERRANEAN SEA

LEBANON

Euphrates River

Tigris River

p.115 Dasht-e-Kavir

IRAN

JORDAN

IRAQ

p.104-5 Dead Sea Rift

Suez Canal

ISRAEL

p.104 Dead Sea

p.100-1 Zagros Mountains

Gulf of Suez

Gulf of Aqaba

p.102-3 Red Sea

p.108-9 Kuwait Oil Fires

Shatt al Arab

Nile River

p.114 An Nafūd

KUWAIT

p.101 Salt Domes

PERSIAN GULF

MIDDLE EAST

EGYPT

BAHRAIN

p.112-14 Najd Plateau

QATAR

Strait of Ho

UNITED ARAB EMIRATES

p.111 Ruб al Khali

SAUDI ARABIA

p.109 Abu Dhabi

SUDAN

RED SEA

no defined boundaries

OMAN

YEMEN

ETHIOPIA

INDIAN OCEAN

DJIBOUTI

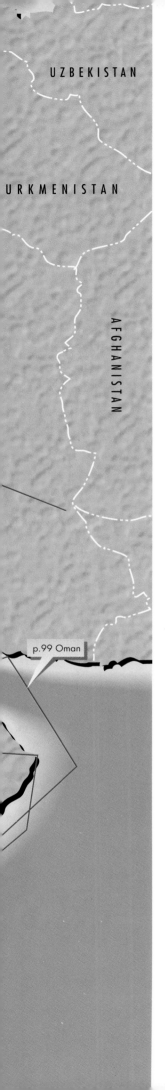

huge sand sea that covers the lower third of the Arabian Peninsula.

A steep scarp faces the Red Sea along the western margin of the peninsula. The mountains along this coast are the highest on the landmass. The Asir Mountains of western Yemen reach elevations greater than 3000 m (9,840 ft.). Southeasterly winds push warm, moist air up the mountain slopes from the coast, and clouds often form above the Asirs. Precipitation from these clouds provides the highlands of western Yemen with enough rainfall to support forests of juniper and acacia. Similar vegetation is found across the Red Sea on the cloud-covered Ethiopian Plateau. (See bottom left of mosaic.)

The rift that separates the African continental plate from the Arabian plate is submerged beneath the Red Sea and the Gulf of Aqaba. To the north of Aqaba the rift is visible, particularly where it contains the Dead Sea and Lake Tiberias (Sea of Galilee). The Red Sea Hills overlook the northern Red Sea and are the major highlands of Egypt. The Sinai Peninsula lies between the Red and Mediterranean seas. The seas are connected by the Suez Canal, which lies along the western margin of the Sinai. Historically and currently the Sinai Peninsula has been both bridge and battlefield between nations, regions, and continents.

On the northern shore of the Mediterranean Sea, Turkey forms a peninsula that lies on a small plate moving to the west between the larger African and Eurasian plates. The landmass consists of a central plateau bounded on the north and south by folded mountain belts that converge in eastern Turkey. Most of Turkey has a semiarid climate with long, cold winters at higher elevations in the east. Basins on the interior Anatolian Plateau contain shallow salt lakes. Tuz Gölü is highlighted by white evaporites surrounding the lake, which is less than a meter (3.28 ft.) deep.

The folded mountain belts that meet in eastern Turkey diverge in Iran, forming the Elburz Mountains, which lie along the southern coast of the Caspian Sea, and also forming the parallel folds of the Zagros Mountains in southern Iran. The highlands of eastern Iran enclose basins that contain broad, flat salt deserts and intermittent salt lakes.

In the mountains of eastern Turkey the Tigris and Euphrates rivers rise and flow across Syria into Iraq, where they form an extensive floodplain. The rivers get close to within 30 km (18.6 mi.) of each other near Baghdad and then diverge to flow along the margins of the river lowlands to meet again near the marshes at the southeastern end of the floodplain. The combined rivers flow to the Persian Gulf through a tidal channel known as the Shatt al Arab (Waterway of the Arabs). Sediments transported by the Shatt al Arab form mud flats at the head of the Persian Gulf, extending from the northern coast of Kuwait to the western shore of Iran.

The Persian Gulf itself is contained in a large basin south of the Zagros Mountains, extending from the Syrian Desert in the west to the narrow Strait of Hormuz in the east. The average depth is less than 37 m (about 120 ft.), with the deepest water near the Strait of Hormuz. Sediments beneath the Persian Gulf that have been accumulating for more than 100 million years contain organic materials that have formed the world's largest known reservoirs of oil. Exploitation of this oil has led to a restructuring of the region's political power.

What the Land Brings Forth

GEOLOGICALLY, THE MIDDLE EAST IS AN ACTIVE REGION whose complex physical structure reflects the stresses and upheavals of the Earth's crust. The interactions here of three major plates and several smaller ones have shaped the geography and topography of the area.

As Africa and Arabia slowly separated, the basins formed between them became the Red Sea and Gulf of Aden, while on the eastern side of the Saudi Arabian Peninsula, the Arabian plate converged on the Eurasian landmass. This collision compressed and folded the land, forming the relatively young, rugged Zagros Mountains, (see pages 100-1). The rift running up the Red Sea is a continuation of the structure that forms the African Rift (see Chapter 1), and it extends northward through the Dead Sea and the Jordan River valley, (see pages 104-5). Here the terrain on one side of the fault may have shifted 100 km (60 mi.) relative to the other side. Earthquakes are common in this region and in other areas along the major Mideast plate boundaries.

Broadly, the Middle East can be divided into a mountainous zone in the north and a lower-lying area of plateaus, plains, and smaller ranges to the south. The northern mountains, young and unweathered, formed during the same mountain-building period that produced the Alps 50 million years ago. The two mountain ranges of northern and southern Turkey come together in the east in a zone of volcanic terrain. Here the snow-covered peak of the inactive volcano Mount Ārārāt is found, (see pages 106-7). The Elburz and Zagros mountains branch out from this region along the northern and southwestern borders of Iran, surrounding a broad plateau.

Mountains in the Mideast's southern region tend to be confined to smaller areas and lower elevations. The Red Sea, for example, has a border of uplands on both sides. To the west, the hills of

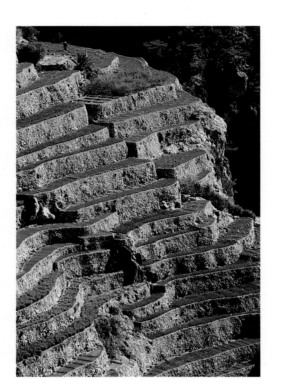

Many areas in the Middle East are fertile because humankind has worked with nature. In the Ḥajar range, which forms the backbone of Oman, the terraces carved into the canyon walls of al Jabal al Akhḍar, the "green mountain" ridge, support spring-fed crops of garlic and onions.

eastern Egypt rise to heights of more than 1800 m (about 6,000 ft.). On the sea's eastern shore, the Hijaz and Asir mountains of Saudi Arabia and Yemen rise steeply from the coast.

Other ranges on the southern edge and southeast corner of the Saudi Arabian Peninsula complete a chain of uplands that borders the stable inland zone. Here sedimentary rocks overlie the ancient crystalline rocks of the Arabian Shield, which was joined to the African landmass before major rifting occurred. In central Saudi Arabia the Najd Plateau slopes gently downward from the western highlands toward the Persian Gulf. It is characterized by parallel scarps formed from limestone which, in very dry climates, resists the forces of weathering. To the south lie the stark tracts of the Empty Quarter.

The geology of the Middle East has determined the region's economic development. In times past, when seas encroached on the terrain, remains of aquatic creatures collected in mud deposits. Later, as the sediments consolidated, increased temperatures and pressures converted the organic matter to hydrocarbons. As the matter was trapped in porous rocks confined between impermeable layers, or collecting against barriers like salt deposits, reservoirs of oil formed. In 1908 the first Middle Eastern oil well began producing crude oil in the Zagros Mountains of Iran. Crude oil output increased in the region during the 1920s and 1930s as wells began producing in Iraq and Saudi Arabia. Oil production sharply increased after World War II to meet the growing demand of industrial powers for petroleum products. In the 1950s exploration to find new oil fields began in the Persian Gulf, and new wells were in production offshore by the 1960s. Such discoveries of huge deposits of oil in the Middle East throughout the 20th Century dramatically changed the economy and lifestyles of many nations and also had far-reaching effects on global economics and politics.

Oman

In this oblique view of the Sultanate of Oman, looking toward the southwest, a line of cloud formations marks a mountain range called al Ḥajar (the Stone), which reaches heights of about 3100 m (more than 10,000 ft.). These clouds form when moist air, such as that blown in over warm seas, is deflected upwards by mountain barriers. At these higher altitudes, the air cools. This accounts for the frequent rain in these mountains, especially at the highest point, al Jabal al Ak-hḍar (Green Mountain) Ridge, a relatively heavily populated and cultivated region.

Oman, on the southeastern corner of the Arabian Peninsula, has a narrow humid coastal plain running along the Gulf of Oman. About a third of Oman's people live in this fertile region, especially around the capital city of Muscat. Inland of the mountains, sandy and stone deserts give way to the Ruḃ al Khali (the Empty Quarter).

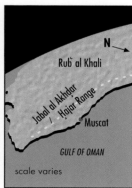

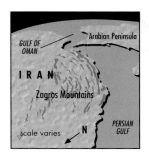

Zagros Mountains

The distinctive Zagros Mountains stretch along the western and southern boundaries of Iran. The central Zagros range (opposite page) with elevations of up to 4000 m (13,000 ft.), borders the Persian Gulf and exhibits a series of parallel ridges and salt domes that are separated by deep valleys. The ridges are folds in the terrain resulting from the collision of the Arabian and Eurasian plates.

The uplands of the central Zagros are home to nomads and their herds.

Salt Domes

The dark, circular features dotting this image are salt domes in the central Zagros range. The blue waters of the Persian Gulf fringe the land. Some of the domes are as high as 1500 m (around 5,000 ft.). They occur where layers of sedimentary rocks overlie thick deposits of salt. The salt, which is less dense than the surrounding rock, flows upward, deforming the sedimentary layers.

In geologically unstable regions, the stresses involved in rock deformation and movement may also squeeze salt upward. When petroleum deposits are present, they are often trapped along the sides of the salt structures. Salt domes are therefore important geologic clues in the search for oil.

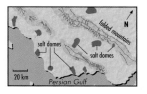

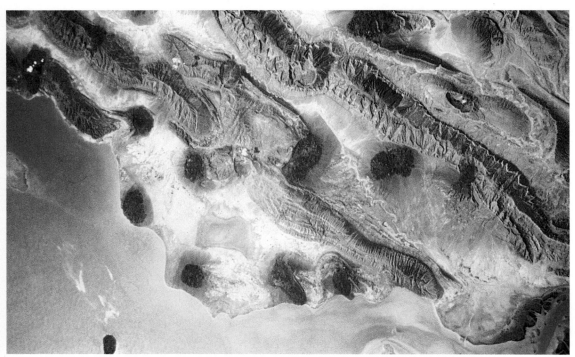

The Red Sea

This oblique view, looking west from the Arabian Peninsula, stretches across the Sinai Peninsula and the Red Sea and into the Eastern and Western deserts of Egypt. Clearly visible is the continuation of the Red Sea rift complex up the Gulf of Aqaba and along the Jordan River valley.

Named for occasional red discolorations of its waters caused by dying algae, the Red Sea is 300 km (180 mi.) at its widest and extends from the region around the Sinai in the north to the narrow strait of Bab el Mandeb near Djibouti in the south. At its northern end the Red Sea branches around the Sinai Peninsula into the fault basins of the Gulf of Suez and the Gulf of Aqaba.

No rivers bring fresh water to the Red Sea, and high rates of evaporation concentrate salts. Pools of exceptionally salty brine also exist in deep parts of the sea, which some scientists believe are fossil waters seeping from ancient sedimentary rock layers. In the Sinai wilderness, a Bedouin girl (opposite top) blends with rocks that dot the landscape.

The Dead Sea

Lying at the lowest point on Earth, the Dead Sea shows itself (opposite page) to be not a sea but a landlocked salt lake. It is located along the Dead Sea rift, which cuts from north to south across this mosaic. The Jordan River flows south through much of the rift valley and empties into the Dead Sea, which lies in a depression formed by what geologists call a down-dropped fault block. Included in the view are Israel, the Sinai, Jordan, and adjacent areas.

The rift zone delineates a boundary between the African and Arabian plates. Evidence exists of complex motion along the faults, with both horizontal and vertical displacements along the line through the Gulf of Aqaba, the Dead Sea, and Lake Tiberias (the biblical Sea of Galilee).

Elsewhere in this image, the coastal cities of Beirut, Haifa, and Tel Aviv stand out in blue colors. Further inland, the city of Damascus is seen as a blue area surrounded by a red-colored vegetated zone.

The Dead Sea is extremely saline and therefore lifeless, except for some species of bacteria. The original source of the salt deposits may have been seas which extended over this region more than 65 million years ago, leaving salt behind as they receded. Today, substantial evaporation helps to concentrate the salts in the Dead Sea so that its waters are seven to eight times saltier than average seawater.

The level of the sea has been falling for decades. Many variables contribute to this shrinking, including climate effects. Stresses of rising population, such as increased drainage of Jordan River waters for irrigation, also lower water levels.

In the years since the imagery for this mosaic was collected, the shallow southern basin of the sea has been separated from the north as the waters dropped around the Lisan Peninsula, which appears as a white triangle jutting into the sea. A scene from 1989 (below) shows the southern basin detached from the northern waters. Here the southern Dead Sea consists of a series of evaporation ponds filled by waters pumped down from the north. The ponds are used to collect potash, a compound used in fertilizers.

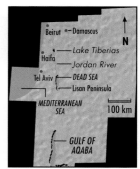

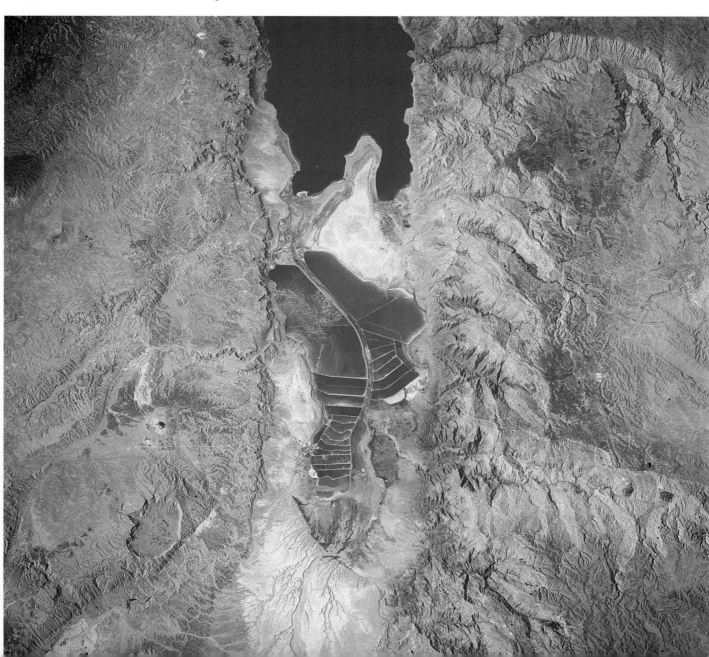

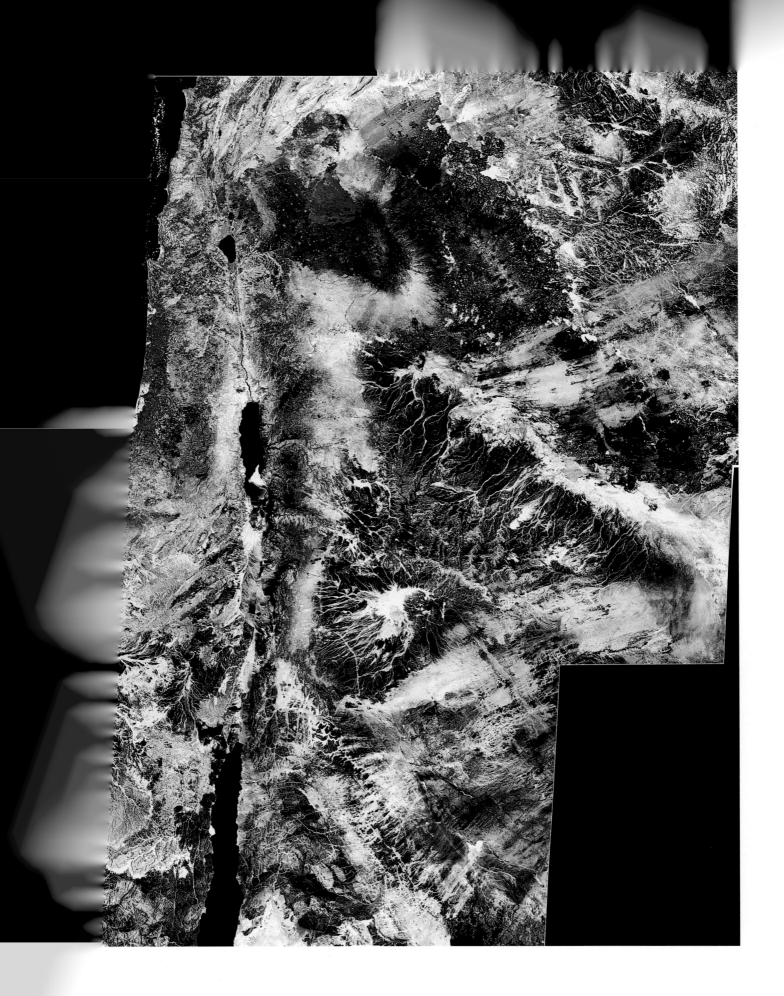

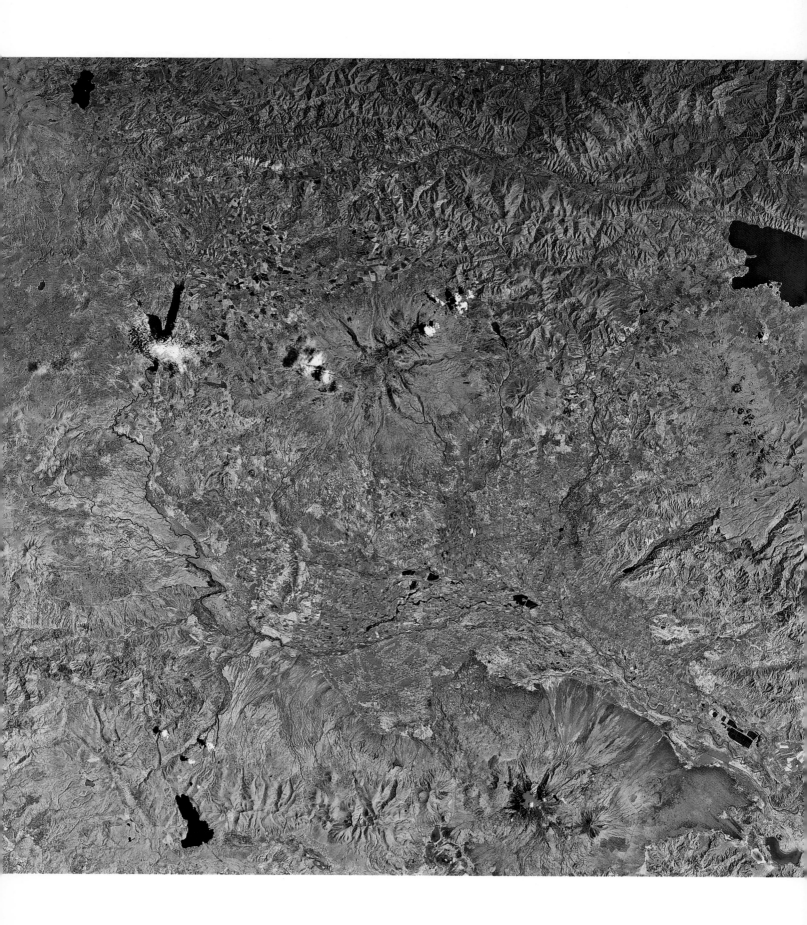

Mount Ārārāt

The computer processing of this Landsat scene (opposite page) causes the snow cap on Mount Ārārāt (lower right) to appear light blue. Ārārāt, or Bü Ağri Daği, with an elevation of 5165 m (16,941 ft.), is the highest mountain in Turkey, shown (left) in its white-capped majesty. Bü Ağri Daği and Küc Ağri Daği (Little Mount Ārārāt, the smaller mountain to the southeast in the image), are inactive volcanoes. Gullies carved by run-off from rain and melting snow descend from the summits of both volcanoes in radial drainage patterns.

The Aras River, coursing north of Mount Ārārāt, forms the border between the eastern provinces of Turkey and Armenia, formerly part of the Soviet Union. The eroded, inactive volcano north of the Aras River (near the center of the image) is Mount Aragats. Small volcanic craters and cones are visible southwest of Lake Sevan.

Because the region is near the boundary of the Eurasian and Arabian plates, earthquakes occur frequently in this part of the Middle East. (About 90 percent of all earthquakes occur near plate boundaries.)

On December 7, 1988, an earthquake measuring 6.5 on the Richter Scale devastated parts of western Armenia. The earthquake was centered at a point on a fault northwest of Spitak. That town and the city of Leninakan (now Gumari) to the west were almost totally destroyed and about 25,000 people were killed. The death toll was largely due to the collapse of poorly designed and poorly built high-rise apartments. There had been historical precedence: a 1926 earthquake ripped through Leninakan, but reports on the extensive damage and loss of life were suppressed by Stalin's government. The Soviets then built low-quality, high-rise apartments despite the region's long earthquake history.

Kūh-e Sahand

In a radar image (below), deep valleys radiate out from the eroded volcanic dome of Kūh-e Sahand, a volcano 3700 m (12,136 ft.) high. The dome was built by repeated eruptions of ash and lava.

Volcanism in this region has resulted from the interaction of three tectonic plates that meet near the border between Turkey and Iran. Kūh-e Sahand lies east of Lake Orūmīyeh, near Iran's border with Turkey and Azerbaijan.

To the west, the Shāhī Peninsula, which also has an eroded volcanic cone, forms part of the eastern shore of Lake Orūmīyeh, the largest perennial lake in Iran. It has an average depth of about 5 m (16.4 ft.) and lacks outlets to the sea. This November 1981 image shows a causeway under construction.

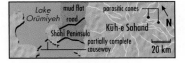

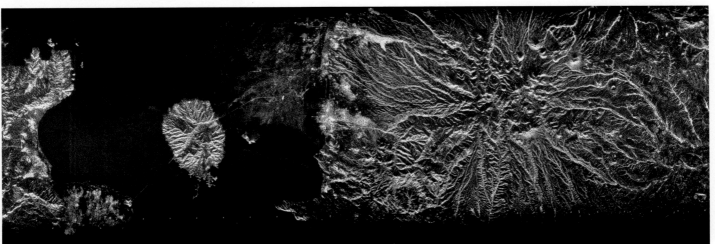

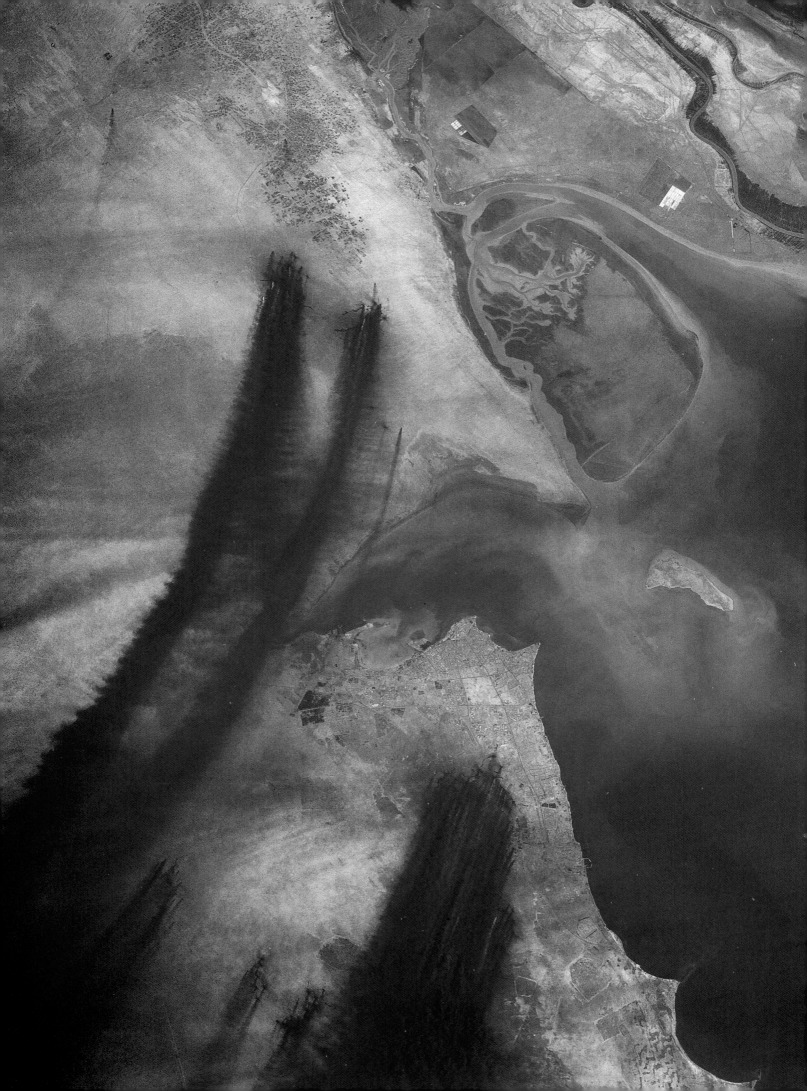

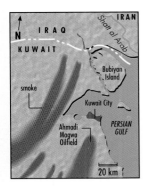

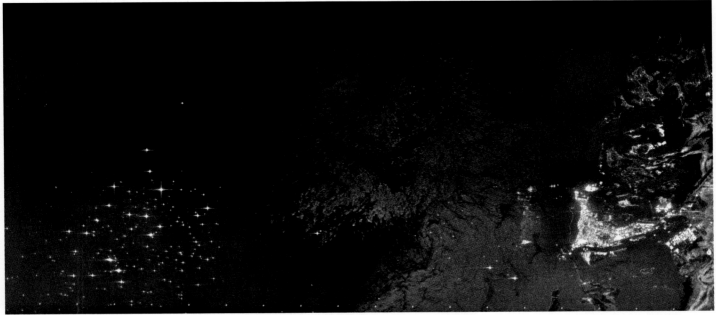

Kuwait Fires

Smoke plumes stripe this image of Kuwait, photographed by astronauts aboard the Space Shuttle in the spring of 1991. As the defeated Iraqi army retreated from Kuwait in February 1991, Iraqi saboteurs blew up more than 700 wellheads, and most of these caught fire.

Dire predictions about the impact of the burning oil on the global climate proved unfounded, but short-term regional and local weather was affected. Black smoke darkened the skies over Kuwait, and black rain fell in parts of Saudi Arabia and Iran.

The top of the smoke plume reached about 5000 m (16,000 ft.), well below the stratosphere. If the smoke had risen higher, climate might have been altered over a much larger area.

Firefighters (left) from around the world came to fight the oil fires. By early November 1991 the last oil fires were extinguished, much earlier than the original estimates of two to three years. Though the fires are out, Kuwait still faces a massive cleanup.

Abu Dhabi

More than half of the world's proven recoverable petroleum reserves are found in the Persian Gulf region. In a radar scene (above), a cluster of bright points at the left is caused by radar reflecting off the steel platforms, derricks, and ships of the Zukum offshore oil field in the southern Persian Gulf. Buildings and streets in the port of Abu Dhabi (Abū Zabī) are radar-reflective over a larger area. Abu Dhabi, on an island off the coast of the United Arab Emirates, is the capital of this nation. Coastal marshes and nearby low islands are less reflective than the city.

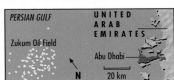

Dunes in a Sea of Sand

EXTREMES OF ARIDITY CAN BE FOUND IN MOST NATIONS OF the Middle East, whereas humid climates occur only in coastal areas of Turkey, the Zagros Mountains region of Iran, and areas of relatively high rainfall along the coasts of the Caspian Sea and the eastern Mediterranean. Much of the Middle East falls within the belt located between 15° and 30° latitude, where warm, dry, subtropical air descends, producing a zone of arid climate. Here are found the Middle Eastern deserts of North Africa, the Arabian Peninsula, and southeastern Iran.

North of this belt lie the arid and semiarid lands of Israel's Negev Desert and the Syrian Desert of northern Saudi Arabia, Syria, Jordan, and Iraq. These deserts are not all sandy: some are covered by lava flows, barren rock, or gravel plains. Great sandy deserts accumulate in lowlands of the Saudi Arabian Peninsula. The three large sand seas, An Nafūd, Ad Dahnā, and Rub al Khali (opposite page) are vast and largely uninhabited regions with broad areas that are unexplored and hard to traverse.

Political boundaries are often ill-defined amidst the barren, largely lifeless tracts of the Empty Quarter. The northern borders of Oman and Yemen, which separate the two nations from Saudi Arabia, for example, are usually not shown on political maps because the boundaries are lost in the wastes of the Rub al Khali. The status of these boundaries remains undefined due to the Rub al Khali's remoteness and the independent nature of the nomadic Bedu, a collection of pastoral tribes that herd camels, sheep, and goats throughout the Arabian Peninsula.

The tribes of the Rub al Khali were among the last Bedu to be affected by 20th–Century economics and materialism. Today many of them have chosen to abandon the relative freedom of nomadic life for employment in Saudi Arabia's numerous oil fields, replacing their camels with pickup trucks.

Rub al Khali, called the Empty Quarter, is a sea of sand the size of Texas, inhabited by few people. The region's shifting dunes and salt flats may hide oil fields that are yet untapped.

The Nafūd in the north (see page 114) is a desert of reddish-colored sands extending across the north side of the Najd Plateau. It merges with Ad Dahnā, which forms a narrow arc outlining the eastern edge of the Najd Plateau. To the south is the forbidding Rub al Khali, or Empty Quarter, which is characterized by sand sheets, saline flats, muddy soils, and a variety of dune forms, some more than 200 m (660 ft.) high.

Dunes of arid lands are shaped by the strength and direction of winds as well as by terrain and amount of available sand. A wide variety of simple and complex forms exists subject to many classifications. Four basic, common dune types are barchan, transverse, longitudinal, and parabolic.

—Barchan dunes are crescent-shaped with a gentle slope on the windward side and a steep slipface on the lee. The points or horns of the crescent point downwind. Barchans form where the wind direction is constant.

—Transverse dunes have a cross-section similar to barchans but are long, linear, and lie perpendicular to the direction of the wind. They are found where sand is abundant and winds are strong.

—Longitudinal, or seif, dunes are also linear in form, but the length of the dune parallels the wind direction instead of cutting across it.

—Parabolic dunes are curved in shape with their horns pointing upwind. They form in lightly vegetated regions.

The dune forms of the Empty Quarter and the other Middle East sand seas merge with one another in complex shapes and patterns that represent the interaction of various winds and terrains. The variety of form and texture, from razor sharp ridges to rolling hills, may seem random but is a direct reflection of the local desert environment. These arid landscapes, though harshly inhospitable to human beings, possess a natural beauty that can be well appreciated from above for both striking appearance and magnitude.

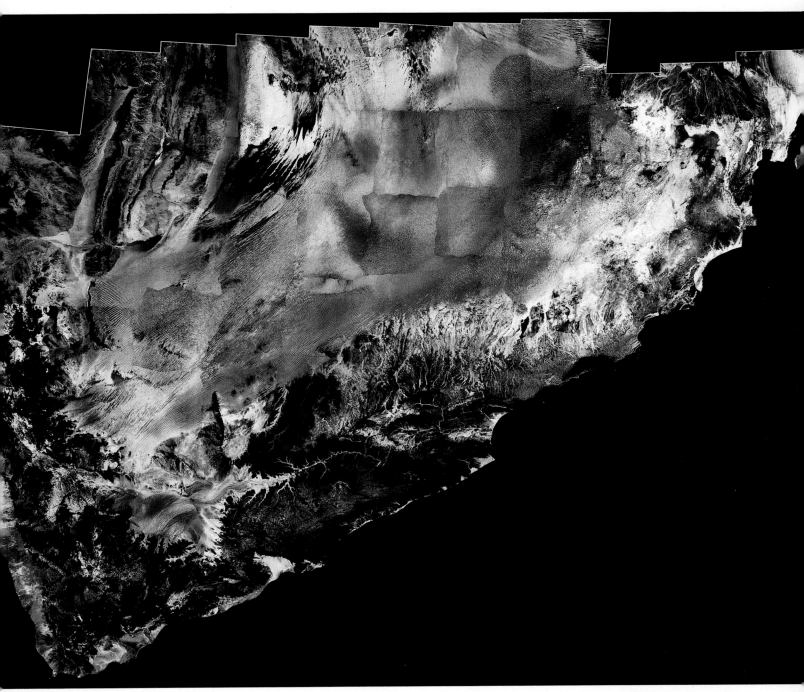

Rub' al Khali

The Arabs of the northern Arabian Peninsula named the great sand sea to the south the Rub' al Khali—the Empty Quarter. This false-color mosaic of more than 50 Landsat scenes suggests why. The region, which its few inhabitants know as the Sands, covers more than 550000 sq. km (214,500 sq. mi.).

The linear dunes in the center and the west seem to advance and grow from northwest to southeast like ocean swells that increase with time and distance under a steady wind.

To the northeast the crescent and star-shaped dunes are more like choppy waves kicked up by a squall. The linear dunes in the southern Rub' al Khali can reach heights of 100 m (328 ft.) and can extend for 200 km (124 mi.). Star dunes at the desert's eastern end can be as high as 150 m (492 ft.).

The blue-gray patches clearly lying between the yellow-orange dunes in the northeast are sabkhas, or salt flats. The broader blue areas outlined by the curving edges of individual scenes are possibly sands that have recently been rained on. In false-color imagery, wet soil appears blue because moisture decreases soil reflectivity. Many of the blue areas adjacent to the Asir Mountains and the Jabal Tuwayq are alluvial gravel deposits.

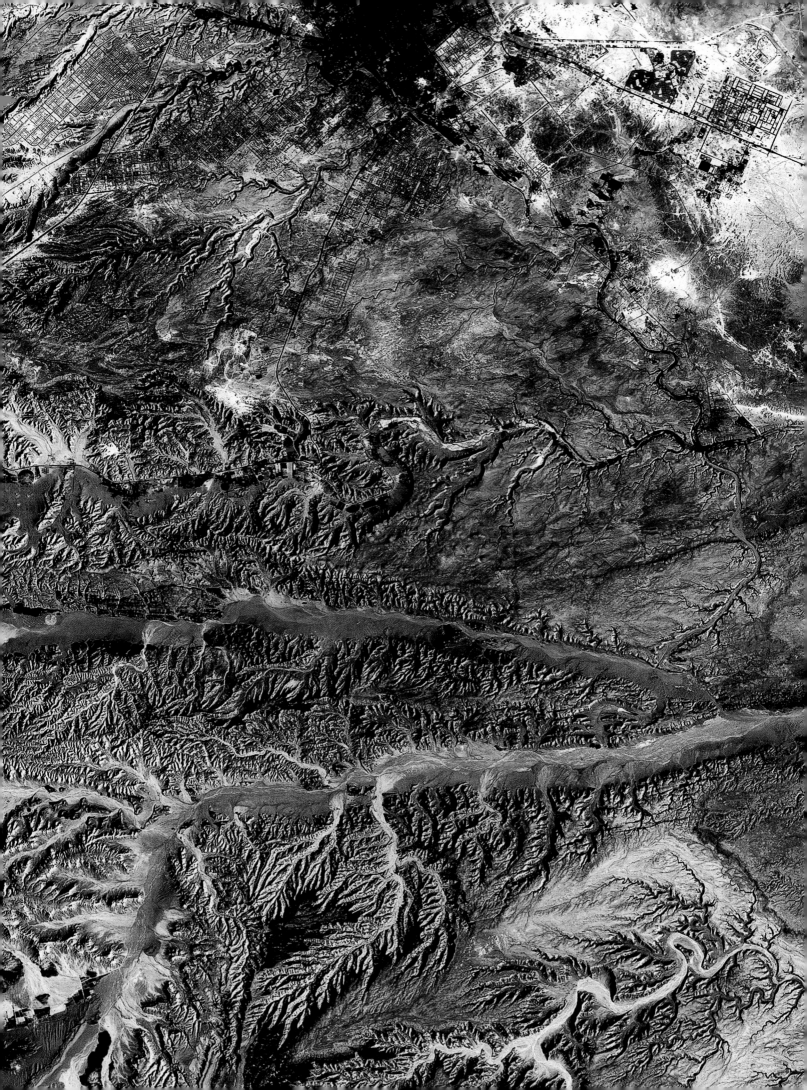

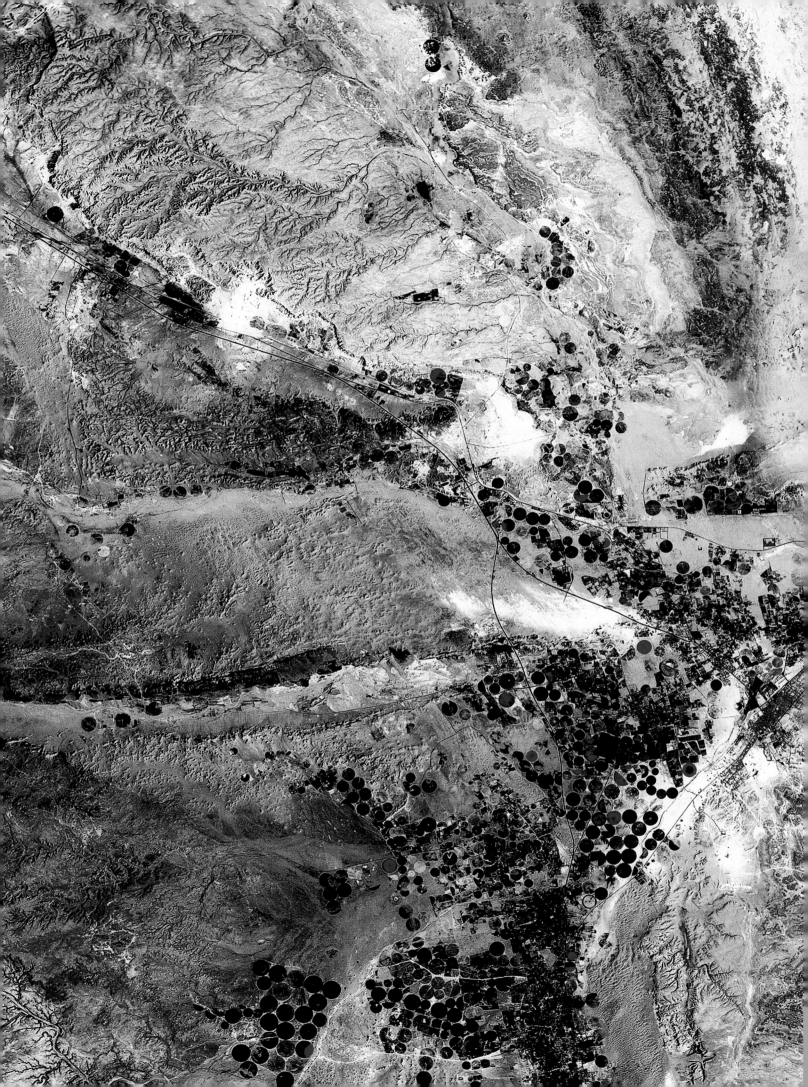

(previous pages)

Najd Plateau

Riyadh (top left in image), the capital of Saudi Arabia, lies near the northeastern edge of the Najd Plateau; here are numerous dry river channels (wadis). A few may carry water after rare torrential rains.

This area, site of several oases, is one of Saudi Arabia's most densely populated regions. Clusters of circles (lower right) are the product of center-pivot irrigation systems. In these systems, sprinklers are attached to a rotating arm anchored in the center of a field.

Many irrigation systems in Saudi Arabia tap the water stored for thousands of years in permeable rock formations lying deep beneath the surface. Vast quantities of this water exist and can irrigate large areas for decades. But the supply is not inexhaustible, and is replenished exceedingly slowly, if at all.

A large network of roads and railroad tracks can be seen extending in every direction from oil-rich Riyadh, which has grown from a village and supply station for desert travelers in the 1930s to a city inhabited by more than 1.5 million people.

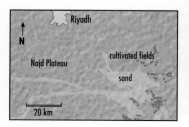

An Nafūd

An Nafūd—meaning simply "the desert"—is the largest of the sand seas in northern Saudi Arabia. Huge dunes in the Nafūd can reach heights of 90 m (roughly 300 ft.).

Although this remote area is hard to traverse, nomads are drawn here with their herds because the light seasonal rains bring short-lived vegetation.

In the image below, masses of transverse and coalesced barchan dunes are seen on the southern edge of the Nafūd.

Jabal Ajā (lower right-hand corner) is a granite plateau. Granitic hills in the region, scattered along the edge of the Nafūd, create "wind shadows" that act as barriers to the continually shifting sands.

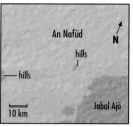

Dasht-e-Kavir

A large salt and gravel desert in eastern Iran, the Dasht-e-Kavir lies in a basin enclosed by several mountain ranges. Dark gravel skirts are visible along the mountains to the north (top left) and south (bottom center). The gravel is deposited by the spring run-off that temporarily collects in broad, shallow lakes in the lower parts of the basin.

When the spring run-off seeps into the basin floor, salty groundwater rises to the surface and evaporates, leaving a white crust. Older salt deposits appear less white because windblown dust collects on the surface. Under the salt crust, in some areas, black mud expands on hot days, cracking the crust, tilting the broken pieces as high as 50 cm (19 in.), and forming a rough black salt desert. White and black salt deserts are visible in the eastern third of this Landsat scene.

The whorls are outcrops of folded sedimentary rock on the flat basin floor. The light-brown and yellow curves are generally the exposed edges of rock layers. Darker curves are created by wet soils lying in shallow depressions between the exposed rock.

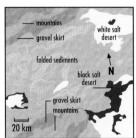

SHAPER OF NATIONS

THE ANCIENT GREEKS BELIEVED THAT THE WORLD HAD AT LEAST THREE continents: Europe, Asia, and Africa. Culturally, Europe does have the unity of a continent, but physically Europe is a collection of islands and peninsulas extending from Eurasia, the Earth's largest landmass.

From the main peninsula that is considered the continent of Europe project the Crimean, Balkan, Italian, Iberian, and Jutland peninsulas. The Scandinavian peninsula, also known as the Fennoscandian Peninsula when Finland is included, connects to Eurasia in northern Russia.

The larger islands associated with Europe include Britain, Ireland, Sicily, Sardinia, Corsica, and Crete. Iceland and the islands located north of the Arctic Circle in the Barents Sea, such as Svalbard and Franz Josef Land (not visible in the mosaic on the opposite page), are also regarded as part of Europe. Iceland, the extreme northwestern part of Europe, has several large permanent ice plateaus and straddles the rift separating the North American plate from the Eurasian plate.

Although Europe's eastern boundary with Asia has many geopolitical definitions, one generally accepted boundary runs along the Ural Mountains and down the Ural River to the Caspian Sea, dividing Russia between Europe and Asia. The crest of the Caucasus Mountains, between the Caspian and Black seas, is usually looked upon as Europe's southeastern border with Asia. Like Russia, Turkey spans the intercontinental boundary.

The youngest mountains in Europe, the snow-covered Alps, are the culmination of a complex mountain-building episode derived from the collision of the African plate with the Eurasian continental plate. Over the last 40 million years large-scale folding, thrusting, and vertical uplift have occurred at this active margin of tectonic plates. Geologic forces in the region moved huge slabs and piled up mountain ranges such as the Alps and Jura Mountains so that the landscape dramatically demonstrates the mobile and dynamic nature of Earth's crust.

In the British Isles and the snow-covered uplands of Scandinavia, the mountains are not as rugged as the Alps and are about 450 million years old. At the southern end of another great European range, the Carpathian Mountains, the Danube River cuts through a gorge called the Iron Gate, an historic trading and invasion route, and then flows across a broad plain to the Black Sea.

In northeastern Europe the advance and retreat of continental glaciers over the past 2 million years has gouged the continent's older, crystalline basement rock. Many of the lakes in northeastern Europe occupy troughs carved by the continental glaciers. Mountain glaciers carved linear valleys, creating the long, narrow lakes

Ural River

KAZAKHSTAN

p.125 Volga River

olga River

CASPIAN
SEA

GEORGIA

AZERBAIJAN
ARMENIA

IRAN

SYRIA

Tigris River

Euphrates River

IRAQ

SAUDI
ARABIA

on the northern and southern flanks of the Alps. Jutting out from continental Europe, the Iberian Peninsula is its farthest extension. Arrayed across the peninsula, isolating it from the rest of Europe, are the Pyrenees Mountains. In the mosaic, a dark triangle north of the Pyrenees indicates the pine forests of the Landes region in southwestern France.

Terrain has helped to shape cultures and nations in and beyond Europe. In some places the land threw up barriers, but in other places rivers and plains offered pathways that opened up the land. Mountain ranges, such as the Alps and the Pyrenees, separated the people of the continent's core from the people of outlying peninsulas.

The steppes of Russia and Ukraine were invasion paths for Hun, Mongol, and other conquerors who swept across Eurasia's grasslands. Similarly, Europe's great rivers provided routes for raiding and trading, allowing Swedish Vikings, or Varangians, to reach the Black Sea. They also established Kiev, a center of trade on the Dnieper River, as the capital of medieval Russia. From Europe's extensive coastline mariners could sail to nearby islands, such as the British Isles and Iceland, and to distant continents where they disseminated their own cultures, artifacts, and languages, and established colonies.

Mineral and energy resources enriched Europe: iron in Sweden and Russia… coal deposits in a basin extending from the British Isles to the Don River in southeastern Russia…oil and natural gas reservoirs beneath the North Sea and in Russia. Rapid industrial development in the 19th and 20th centuries depleted some of these resources and severely damaged the environment.

To cope with continent-wide environmental problems, the European Community has been establishing standards to reduce pollution in its member states. The liberation of Eastern Europe and the demise of the Soviet Union in 1991 has revealed severe damage to water quality, forests, and human health: problems that were ignored by the now-defunct regimes. In the worst areas of eastern Europe tens of thousands of children have bronchial or other respiratory diseases. Some Czech cities are so polluted that school children are moved out each year to spend several weeks in areas where the air is cleaner.

Modern farmers' intensive use of pesticides and fertilizers also contributes to Europe's environmental woes. The chemicals empty into rivers and lakes, adding poisons and nutrients—a double-barreled burden—to the water. The pesticides kill some animals and plants directly; the nutrients drastically alter the balance of life and indirectly doom other forms of life.

More environmental hazards have come from the shipping of petroleum products and offshore oil and gas exploration, especially in the North Sea. Tanker accidents have ruined beaches and wiped out colonies of sea animals, from limpets and oysters to fish and cormorants. In March 1978 the American-owned *Amoco Cadiz*, after losing her steering, drifted onto shoals off the coast of Brittany, dumping about 69 million gallons of oil into the sea. Northwesterly winds pushed the oil onto the shore, blackening about 100 miles of beaches in what was history's largest oil spill. On the shore people put up signs that said: "La mer est morte—the sea is dead." It took more than a decade of clean-up work and natural processes to restore the coastal region of France devastated by the spill.

Always Close to the Sea

EUROPE'S MANY PENINSULAS MAKE IT UNIQUE. BECAUSE these narrow arms give the small continent an exceptionally long coastline, seas reach deep inland and have moderating effects on climate. In Western Europe the sea is rarely more than a few hundred kilometers away from anywhere and has traditionally been a source of food, a means of commerce, and a route for trade and conquest. Europe's many seas have greatly influenced the history and cultures of European peoples.

Draining into the seas is a vast network of European rivers, many fed by the snows and glacial meltwaters of the Alps. The Rhine, the most important European waterway for industrial transport, rises in the Swiss Alps and flows north through Germany, serving farms and industries as a highway of commerce before joining the North Sea near Rotterdam.

The Rhône also originates in the glacial highlands of Switzerland. Moving south across France to the Mediterranean Sea, the Rhône and its tributaries are harnessed by a series of dams for flood control, power, and water storage. Like the Rhine Valley, the fertile Rhône Valley sustains numerous cities, farms, and orchards. The rugged Rhône is not as useful for transport as the Rhine. But, as a land route, the Rhône Valley provides the only easy access from the Mediterranean coast to central France.

The Italian Alps give rise to the Po River, which travels through northern Italy to the Adriatic Sea (see page 142). The Volga, Russia's major waterway (see page 125), and the Danube, which cuts its way across mountain ranges and traverses eight countries on its route from Germany to the Black Sea, are Europe's longest rivers.

Although the rivers of Europe have served traditionally as political boundaries and sources of food, on this highly industrialized continent they are most heavily used as commercial arteries. Canal systems in both western and eastern Europe link major rivers, tying together the distant industrial centers and providing water highways for inland shipping and access to the sea.

Land's End (above) is the rocky point of Cornwall that meets the sea at England's southwest extremity. For centuries, Cornwall's treacherous coastline has caused countless shipwrecks.

With the exception of the Scandinavian highlands, Europe's mountain ranges are relatively young geologically and are somewhat low-lying in comparison to the great ranges of Asia and the Americas. The Alps (opposite page) display Europe's highest, most rugged peaks, carved by extensive glaciation. Run-off from melting alpine snow feeds several major rivers, which course through glacially sculpted valleys before departing for lower terrain.

The Alps are part of Europe's south-central highlands, a belt of ranges (including the Pyrenees and the Carpathians) stretching from one side of the continent to the other. This band of folded mountains developed partly in response to pressures from the northward movement of the African continental plate.

Two other groups of ranges rise on the edges of the continent. The northern highlands (see pages 128-29) lie in Scandinavia and the British Isles. The Urals, rich in minerals, reach heights of only 1900 m (6,200 ft.). They are often chosen as the eastern boundary of Europe.

There are few volcanoes in Europe; volcanic activity is centered mainly in Iceland (see page 129) and the Apennine region of Italy. A string of volcanic lakes and mountains, including Vesuvius and Mount Etna, extends down the west coast of Italy into Sicily. Etna, which reaches a height of more than 3 km (nearly 2 mi.) and ranges from 40 to 60 km (25 to 37 mi.) in diameter, is the largest active volcano in Europe. The earliest written report of an Etna eruption came in 693 BC, and lava eruptions continue into the present. In spite of the volcano's persistent activity, these slopes are home to one of the world's most densely concentrated farm populations.

Europe's three major mountain regions border coastal and central lowlands. Broken only by rolling hills, these flatlands constitute one of the largest regions of level plains on Earth.

The Alps

The snow-covered Alps of southern Europe highlight this Space Shuttle photograph. Monte Rosa (4635 m or 15,200 ft.), the second-highest peak in the Alps, is visible, as are several other prominent peaks, including the Matterhorn, the Finsteraarhorn, and Jungfrau. (Mont Blanc, the highest, is just off the image.) Valley glaciers surround the taller peaks. The Jura Mountains are visible north-northwest of the Alps.

Many valleys and troughs in the Alps have been sculpted by the intense erosion of mountain glaciers. Bowl-shaped basins called cirques, knife-like ridges called arêtes, and sharp-peaked horns display some features produced by abrasion, ice wedging, and glacial plucking. Most of the lakes visible in the image have a glacial origin. (D.M.H.)

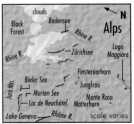

Crete

About two-thirds of Crete, the largest of the Greek islands, is visible in this Space Shuttle photograph. Crete, which marks the boundary between the Aegean Sea and the Mediterranean Sea, is dominated by a harsh, mountainous terrain. Mount Ida (2460 m or 8,070 ft.) in the Psiloriti Mountains and Lefka Ori (2450 m or 8,040 ft.) in the White Mountains are the highest peaks.

The island of Crete, as well as mainland Greece and the nearby Cyclades Islands, was

the site of a Bronze Age culture called Minoan (after King Minos) that existed from 2500-1000 BC. The Minoans built extravagant palaces and structures on Crete, most notably at Knossos. Kommos (left) on Crete's southern shore, has revealed Minoan treasures to excavators. After the Minoans, strategically important Crete was successively dominated by Dorians, Romans, Byzantines, Arabs, Venetians, and Ottoman Turks. Crete, or Kriti in Greek, was united with Greece in 1913.

The island consists mainly of limestone and dolomite eroded into sharp ridges and cliffs. These rocks may be slowly dissolved by surface water and groundwater, producing a terrain of numerous shallow caves and depressions or sinkholes, known as karst topography. The geologic history of the island is related to the same mountain-building that created the Alps, a complex collision between the Eurasian and African continental plates, as well as several smaller tectonic plates.

As a result of the collision, huge crustal slabs were transported northward as thrust sheets or large overturned folds. The interaction of these slabs at plate margins forms the framework of the geologic structure of Crete. (D.M.H.)

Sicily

This high, oblique view of Sicily, looking south along the "toe" of Italy, shows the plume of Mount Etna blowing eastward over the Ionian Sea.

The largest island in the Mediterranean Sea, Sicily is separated from Italy by the narrow Strait of Messina, clearly seen in the image. The triangular-shaped island has an active seismic zone along its east coast.

Etna lies in this region that has been plagued by earthquakes and volcanic eruptions throughout history. More than 200 small cones lie on the flanks of Mount Etna. One such cone was the source of a 1669 eruption, which was the most devastating episode of Etna's recorded history. The volcanic activity persisted for more than four months, and the nearby town of Catania was virtually destroyed.

Because of its central location in the Mediterranean Sea, Sicily has a rich and ancient heritage. The earliest known inhabitants date back as far as 10,000 years ago. Phoenicians and Greeks settled in the 8th Century BC. Throughout history the island has been a crossroads for settlers, traders, and conquerors. Today, along with some nearby islands, Sicily makes up an autonomous subdivision of Italy.

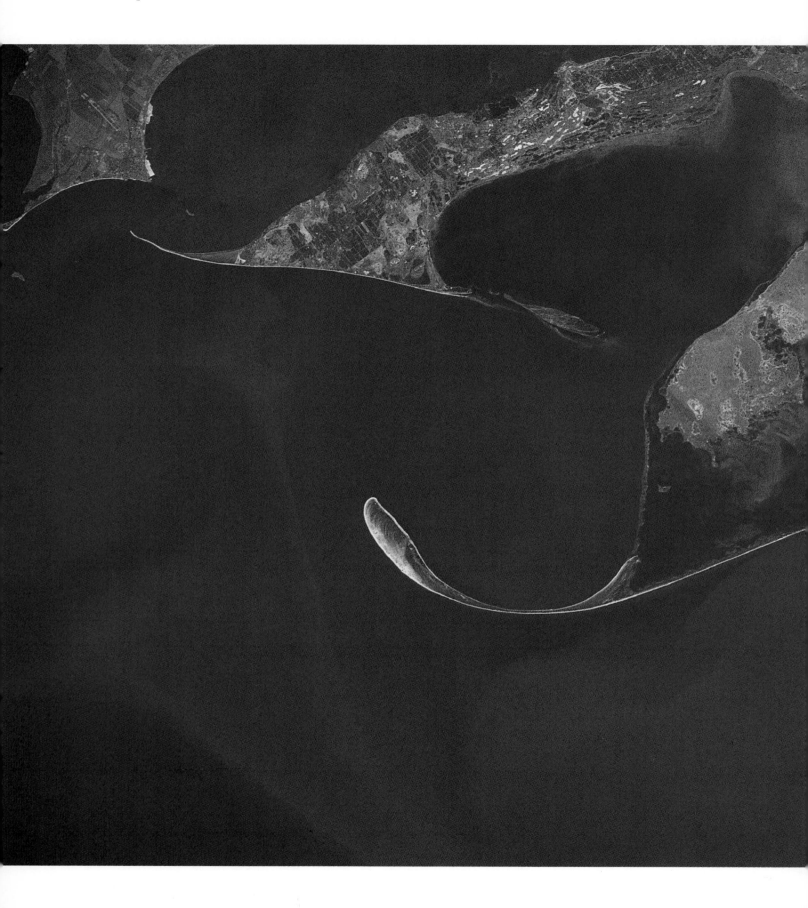

Ochakov

BLACK SEA   10 km

The Black Sea

A large inland sea more than 1200 km (740 mi.) wide, the Black Sea borders on Ukraine, Russia, Georgia, Turkey, Bulgaria, and Romania. Its name may have come from the ancient Turks who found its stormy waters uninviting and hard to traverse.

Joined with the Caspian and Aral seas in one large ocean more than 40 million years ago, the Black Sea was isolated when geologic forces uplifted terrain to form separated basins. Today only the very narrow Bosporus Strait connects the Black Sea to the Mediterranean by way of the Sea of Marmara and the Dardanelles.

The Black Sea reaches depths of more than 2200 m (7,200 ft.). Along the sea floor rests a layer of lifeless water where only hardy bacteria survive. The rain- and river-fed surface waters do not mix with the denser bottom waters. In those stagnant, submerged layers, organic matter collects and decomposes, producing large quantities of carbon dioxide, hydrogen sulfide, and ammonia.

In this scene along the north coast, a long, curved sandy spit has been formed by the interplay of currents and sands. The urban area with piers and an airport along the north coast is a Ukrainian town named Ochakov, whose history dates back to 6th Century BC as a Greek colonial city, Alektor.

The Volga

In an image spanning the horizon, the Volga, one of Europe's mightiest rivers, empties into the cloud-streaked Caspian Sea. Near the apex of the Volga's broad delta is the city of Astrakhan, a port famous for its caviar industry. Perched on islands in the low-lying delta, the city has numerous bridges. Flood-control dikes extend for 75 km (47 mi.) within its boundaries. Ten-inch-wide pink lotus blossoms (above) bloom on the Volga delta near the city.

The Volga flows for 3700 km (2,300 mi.) across what was the Soviet Union. Innumerable dams, reservoirs, and canals along its length reflect Soviet efforts at flood control, power generation, and commercial transport. Canals join the Volga to the Baltic Sea, the White Sea, the Sea of Asov, the Black Sea, and Moscow.

Beyond the image is Volgograd, a major city along the Volga. An industrial and oil-producing center, the city lies about 450 km (280 mi.) from the river's mouth. Formerly known as Stalingrad, Volgograd was the site of a decisive battle in World War II.

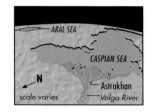

ARAL SEA

CASPIAN SEA

N
scale varies

Astrakhan
Volga River

The Seine

Looping through Paris in this SPOT image, the Seine River paints the City of Light and its environs in shades of blue. The Seine rises in east-central France near the city of Dijon, snakes northwest across the country for more than 770 km (480 mi.), and ends in the Baie de la Seine along the edge of the English Channel near Le Havre. One of the largest rivers in France, the Seine is the country's busiest inland shipping route. It is also the major watercourse in a network of rivers and river valleys that converge in a geological depression known as the Paris Basin.

The geography of the region, reflected in the confluence of these natural trade and transport routes, greatly influenced the location of a city founded on an island in the Seine more than 2,000 years ago by a Gallic tribe known as the Parisii. First known in recorded history as Lutetia (meaning "dwelling in the midst of the water"), this small town grew into the great city of Paris.

Today the Seine travels nearly 13 km (8 mi.) through the Parisian metropolis and is an integral part of the city and its charm. Flowing slowly and rarely flooding, the river has throughout history demarcated differing social and cultural areas of the left and right banks.

More than thirty bridges cross the Seine in Paris. The oldest is the Pont Neuf dating back to the end of the 16th Century. Still strong and sturdy, the bridge runs across the western edge of the Île de la Cité, the island on which the city was first founded.

The Île de la Cité lies in the river at the center of the image. The white shape of Notre Dame Cathedral is on the southeast end of the island. Built from 1163 through 1345, the cathedral of Notre Dame (below) occupies the site where a Roman temple once stood. It was heavily damaged by the mobs of the French Revolution, rescued from sale to a stone merchant by Napoleon Bonaparte (who held his coronation there in 1804), and extensively restored in the mid-1800s.

The Arc de Triomphe is set off by 12 radiating boulevards. Extending southeast from the Arc de Triomphe is the broad, heavily trafficked Avenue des Champs-Élysées.

On a line extending from the Champs-Élysées stands the Louvre Museum, the centuries-old palace that displays such treasures as the Mona Lisa and the Venus de Milo. The Eiffel Tower, casting a long shadow to the northwest, rises to a height of 320 m (1,050 ft.). To the southeast of the tower is the École Militaire across the Champs de Mars. The arc-shaped building behind the École Militaire is the home of the United Nations Educational, Scientific and Cultural Organization. The runways of three airports also can be seen: on the upper right, de Gaulle Airport; just below, Le Bourget Airport; at the lower edge, Orly Airport.

Far to the left, on the outskirts of Paris, lies Versailles, site of the opulent palace built by Louis XIV. The vast complex is large enough to be seen in the image. Clearly visible are the cross-shaped Grand Canal and the Pièce d'Eau des Suisses, the largest of the innumerable decorative pools of Versailles.

British Isles

Surrounded by shallow coastal waters, the British Isles (opposite page) lie within Europe's continental shelf. The coast of France near the port city of Calais (bottom right) is visible across the 33-km (21-mi.) Strait of Dover. The British Isles include two large islands: Britain, which encompasses England, Scotland, and Wales; and Ireland, which includes the Republic of Ireland and Northern Ireland, the latter a part of the United Kingdom. Also in the UK are a number of smaller islands, such as the Isle of Man, the Hebrides, and the Shetland Islands.

Glaciers left their mark on Scotland (top center), where ice-scoured valleys, or glens, are aligned with faults in the highlands. Freshwater lochs have filled parts of some glens,

and from the western coast of Scotland long, narrow sea lochs extend inland. The narrow bays on the southwestern coast of Ireland appear similar to the sea lochs of Scotland, but their origins are not glacial. The southwest coast of Ireland is gradually submerging and the bays, or rias, fill the river valleys between parallel ridges perpendicular to the coast.

On the eastern and southeastern shore of Britain, coastal marshes and mud flats appear dark gray, particularly near the mouth of the Thames estuary and in the Wash. The Thames estuary and river point to the center of the gray mass that is the greater London urban area. On the west coast, Morecambe Bay is highlighted by a light-gray band of tidal flats.

Iceland

Vatnajökull, the largest glacier in Europe, crowns southeastern Iceland (below). In this rugged land, volcanoes and lava flows, hot springs and geysers coexist with icy tracts of glaciers. Iceland's location

just south of the Arctic Circle means its winter days see only a few hours of twilight, while from spring to late summer, darkness does not fall. The population of nearly 250,000 is concentrated in about one-fifth of the land area, with almost 150,000 people living in and around the capital of Reykjavík (not shown).

The great ice cap is about 900 m (roughly 3,000 ft.) thick. Numerous outlet glaciers can be seen flowing from the cap. In the left center, the Grimsvötn caldera marks the site of a subglacial volcano. Geothermal activity under the ice sometimes causes catastrophic flooding in this area.

Scene of frequent volcanic eruptions and earthquakes, Iceland has more than 200 volcanoes, a result of its location along the mid-Atlantic ridge.

(following pages)

Northern Europe

Some of the northernmost lands of Europe appear in icy splendor in this view highlighting the varied terrains of the Scandinavian Peninsula. Mountainous Norway's jagged coastlines are crowded with fjords, deep valleys carved by glacial ice that are now filled by water from the sea. Finland and much of Sweden are relatively low-lying with an abundance of forests and lakes. Although the Scandinavian countries cross the Arctic Circle, their winter temperatures are moderated by the influence of the warm North Atlantic current.

The Baltic, a broad inland sea, borders many nations of northern Europe. The world's largest brackish body of water, it is fed by the fresh waters of hundreds of rivers. The Baltic's salinity remains low because little mixing can occur through the narrow straits leading to the saltier North Sea.

In this March image, sea ice is visible, especially in the Gulf of Bothnia, which lies between Sweden and Finland. Ice often interrupts shipping in the north end of the gulf for more than half the year.

European Commerce

TRADE AMONG THE REGIONS OF EUROPE WAS BEGINNING AS the Stone Age was ending. Along the Mediterranean and Atlantic coasts, early Europeans sought tin and copper for their Bronze Age industries. During this prehistoric period the Indo-European language arrived with migrants from southwestern Asia. The evolution of Indo-European into dozens of distinct languages led to the emergence of as many nationalities. As these nationalities dealt with each other, sometimes in commerce and sometimes in conflict, they shaped Europe's history.

Marseille (opposite), which dates back to at least 600 BC, was a commercial port and an outpost of trade and exploration in the western Mediterranean. Greek colonists founded the city, which they called Massalia. In 330 BC a Massalian named Pytheas, seeking sea routes, reached the Atlantic and claimed to sail beyond Iceland. Greeks colonized what is now Italy and southern France, and trade brought amber from Scandinavia, copper from central Europe, and tin from southwestern England to the Mediterranean coast. Later, intra-continental trade continued to thrive under the protection of the Roman Empire.

Commerce suffered with the collapse of the empire, but in the turbulent centuries that followed, European merchants maintained trade with Africa, India, and China. Because the relatively small amounts of silk and spices that traveled overland from Asia actually increased demand, European desire for these luxuries set in motion the global exploration and colonization that began in the late 15th Century.

Spain, Portugal, the Netherlands, Great Britain, and France competed for raw materials and markets and established empires. At its zenith in the early 20th Century the British Empire controlled approximately one-sixth of the world's landmass.

As competition increased, nations closed their domestic and colonial markets to goods manufactured by rival nations. In the early 19th Century,

In Marseille the Vieux, or Old Port, has a long association with seaborne commerce. Cargo vessels have outgrown the quays of the Old Port, which now harbors an international collection of sail and motor yachts.

for example, Napoleon closed all European ports under his control to British goods. The British countered by blockading the continent. The percentage of total British exports shipped to Europe dropped from 85 percent to 30 percent during Napoleon's rule, and Britain's economy suffered.

Empires were doomed by nationalist movements in colonies pushing for independence from European rule. By the middle of the 20th Century, in the wake of two devastating wars, the great empires had been given up. Trade with their former colonies, now independent nations, remains an important part of European commerce.

After World War II, two very different economic systems were at work on the continent. In the east was a system based on a centrally planned model originated by the Communist government in the Soviet Union. In the west capitalism remained the dominant economic system, though many large industries were nationalized.

Many European economists realized that the relatively small national economies of Europe could not compete with larger economic powers such as the United States. The Common Market was formed in 1958 after the removal of inefficient customs barriers between six European nations. The Common Market became the European Community, and membership has increased to twelve nations. Among the goals of the European Community are political and economic integration.

Economic unification is essential to Europe, since the volume of trade with Pacific nations is now larger than transatlantic trade. Competition with Japan, China, and the other developing nations of Asia has become a major concern in Europe. Political unification now being discussed may make national priorities secondary to the good of the European Community.

The greatest challenge now is the integration of Europe's former Communist nations into the continent's economy.

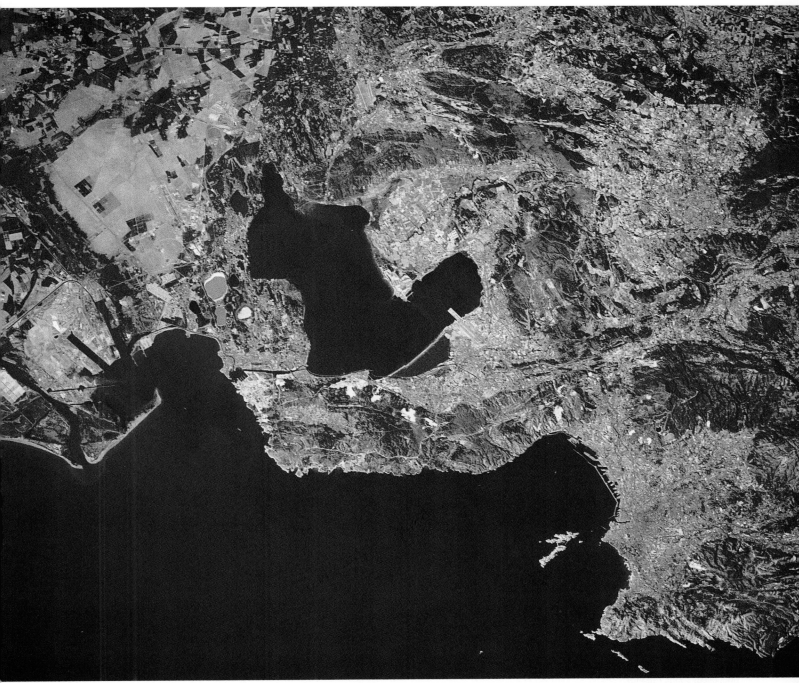

Marseille

The great port of Marseille sprawls across a wide expanse of the Mediterranean Sea. The site of the original city, which was founded about 600 BC, is the Vieux, or Old Port, at the lower right. To the far left, the Rhône River flows past Fos, where a new port was built following World War II after the original port was destroyed by the German Army as it withdrew from Marseille in 1944.

Facilities at Fos accommodate large, deep-draft ships, such as supertankers. An integrated steel mill here gets its raw material by sea, taking advantage of an inexpensive form of bulk transport. The new port also has petroleum refineries and a major pipeline terminal. Lengthened runways at the Marignane Airport extend into the Étang de Berre, a lagoon with a narrow outlet to the sea.

Marseille deteriorated after the fall of the Roman Empire, but revived in the Middle Ages when the city and its surrounding region of Provence became part of France. In the 19th Century Marseille grew as trade with French overseas colonies increased, particularly after the opening of the Suez Canal in 1869. Postwar development of Marseille has improved its economic stature in France, which has traditionally been dominated by Paris and the industrial north.

Strait of Gibraltar

Cloud-streaked Morocco and unclouded Spain nearly touch in this view of the Strait of Gibraltar, which connects the Atlantic Ocean with the Mediterranean Sea and separates Europe from Africa. The strait takes its name from the famous Rock of Gibraltar, which dominates the narrow peninsula jutting from the Spanish coast. In antiquity, the distinctive rock and a point on the southern side of the strait near Ceuta were known as the Pillars of Hercules. In 711 AD Moors crossed the strait from North Africa, bringing Islam and the Arabic language to the Iberian Peninsula. At that time the 426-m (1,397-ft.) limestone rock was named Jabal Tāriq, the Mountain of Tāriq, after a Moorish leader. Over the centuries the name evolved into Gibraltar.

Northwest of the strait, in the upper center of the image, the Rio Guadalquivir empties into the sea. Near its mouth is the port of Cadiz. The city of Seville, one of Spain's largest cities, shows as a gray area about 75 km (46 mi.) upstream.

Seville was a major base for the exploration and exploitation of Spain's colonies in the Americas and the Pacific. In September 1515 five ships under the command of Ferdinand Magellan departed Sanlúcar de Barrameda, at the mouth of the Rio Quadrameda, to circumnavigate the world.

Four ships were lost and Magellan was killed in the Philippines. Almost three years later the remaining ship returned to Sanlúcar, successfully completing the first voyage around the world. Columbus outfitted his three ships at the small port of Palos (upper left) for his voyage across the Atlantic in 1492.

In the early 18th Century the British Royal Navy established a naval base at Gibraltar to ensure access through the strait to the Mediterranean Sea for British ships. Ships could only enter and exit the sea through the strait until the construction of the Suez Canal (1859-69), which connected the Mediterranean with the

Indian Ocean by way of the Red Sea.

With the completion of the canal, the Strait of Gibraltar and the Mediterranean became part of a seaway connecting the maritime nations of western Europe with their colonies in Asia and East Africa, avoiding the longer passage around Cape of Good Hope.

Barcelona

Barcelona's Roman heritage of grid streets shows clearly in this image, where Spain's second-largest city and its environs appear in blue. Barcelona occupies the plain that is north of the mouth of the Rio Llobregat on the Iberian Peninsula's northeastern coast. It is an important Mediterranean port, although the city originally lacked a protected natural harbor. Lengthy moles, or breakwaters, visible in the image, have been constructed over the centuries to keep the Mediterranean's swells from disturbing the port's waters.

Barcelona was probably established as a commercial port by Carthaginian or Phoenician seafarers. After Carthage, a city in North Africa and Rome's rival, sent an army to occupy the Iberian Peninsula in 237 BC, Hannibal, a famous Carthaginian general, marched on Rome, having reinforced his army with mercenaries from Iberia. But the Romans drove out the Carthaginians and then colonized Iberia, leaving their mark on the landscape.

Barcelona's grid has its origins with Roman design, a square plan with north-south and east-west axes possibly based on Roman cosmology, which divided the world into quarters. In Barcelona the axes appear both perpendicular and parallel to the coast.

With a distinct linguistic and cultural identity, Barcelona is the capital of Catalonia, whose language, Catalan, differs distinctly from Spanish. Though the city has a history of separatism, it was the republican stronghold during the Spanish Civil War (1936-39) and had its language and literature banned and suppressed until Spain's liberalization, which began in 1975. Barcelona's harbor (below) is one of the Mediterranéan's busiest ports.

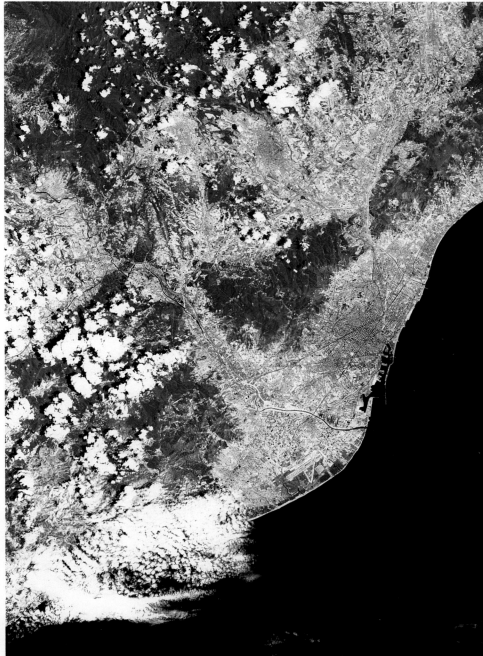

St. Petersburg

Murky waters fringe the shore around St. Petersburg in this Soviet satellite photograph. The thin arms dividing the waters are part of a massive concrete and steel flood barrier that connects Kotlin Island with the north and south shores of the Gulf of Finland.

The barrier, begun in the late 1970s, was designed to protect the city from destructive waves that occasionally sweep up the gulf. But project planners paid little heed to the environmental consequences of the barrier.

Because most of the city's sewage and industrial waste is dumped untreated into the Neva River, the barrier may trap these pollutants in the eastern end of the gulf, creating a foul pool. With the arrival of glasnost in the Soviet Union, people voiced opposition to the project. A fledgling environmental movement succeeded in halting the project in the late 1980s, and some groups are calling for the removal of the barrier.

Czar Peter the Great founded St. Petersburg in 1703. He chose an unlikely site in the marshes at the eastern end of the Gulf of Finland near the mouths of the Neva River. Though the terrain seemed unsuitable, St. Petersburg was strategically located. At the time, Russia's other ports were either on the remote Arctic coast or on the Black Sea, which was isolated from the Mediterranean by the Turkish-controlled Bosporous and Dardanelles straits. The new city and port, with access to the Atlantic via the Baltic Sea, was to be Peter the Great's "window on Europe."

St. Petersburg, flourishing as the Russian Empire expanded across Eurasia, outgrew the banks of the Neva (right). During the late 19th Century, new facilities were built on islands southwest of the city. The Bolshevik revolution of October 1917 began here with a shot fired from the cruiser *Aurora*, which was moored along the north bank of the Neva River, signaling the Bolsheviks to storm the Winter Palace on the Neva's south bank.

During World War I the city's name became Petrograd because its original name sounded too German. After Lenin's death in 1924 the name was changed to Leningrad, and the city continued to develop as an important industrial and shipbuilding center and port. Leningrad's five shipyards produced major merchant and naval vessels, including nuclear-powered icebreakers and deep-diving, titanium-hulled submarines.

In 1991, with the collapse of the Soviet Union, the city returned to its original name. Today, St. Petersburg is the second-largest city in Russia.

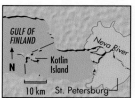

Europe at Night

Lights garland Europe in this image acquired during night passes over the continent. The image, collected by a sensor detecting visible light, dramatically displays the distribution of the continent's people. Most noticeable is the coalescence of cities (left) in northwestern Europe, which includes the northern French city of Lille, the Belgian cities of Brussels and Antwerp, Rotterdam and Amsterdam in the Netherlands, and the German cities clustered in the Rhine and Ruhr valleys. Paris, the City of Light, is the large single bright spot immediately to the south of these areas. London and the cities of the Midlands occupy a large area of southern England.

These three points—Paris, the Ruhr, and the Midlands of England—define the original industrial core of western Europe, the most densely populated and industrialized part of the European continent.

Beyond the industrial core, the rest of Europe's urban population is spread throughout smaller cities that are national and regional capitals and industrial centers. The lights of smaller coastal cities and towns reveal the familiar outline of the Iberian and Italian peninsulas (lower left). Moscow (upper center) visibly dominates the European part of Russia. A network of secondary cities and towns can be seen radiating out from Moscow. St. Petersburg (left) is also visible in the image. The bright points in the North Sea are produced by gas flaring off above offshore platforms that are used for oil and gas drilling and production.

Berlin

The Berlin Wall snakes diagonally across the sprawling urban landscape in this SPOT image of the Wall as a memory. The image was made before the momentous events that began in the summer of 1989, when thousands of East German refugees began fleeing to the West through Hungary. At a massive demonstration in early November of that year, more than 1 million East Germans protested in East Berlin against their government. Three days later the repressive regime of the Socialist Unity Party resigned. On November 9, 1989, travel restrictions between East and West Berlin were lifted. That night millions of Berliners met along the Berlin Wall and began tearing it down (below). The destruction of the wall set in motion the reunification of Berlin and Germany and symbolized the end of the Cold War.

Berlin, originally a medieval market town, was built on the banks of the Spree River (entering the image from the lower right) near its junction with the Havel River. The rivers, tributaries of the Elbe River, were important trading routes between the coast and the lowlands. Grain, furs, timber, pitch, and tar were shipped downstream from the east, and wine, metal, cloth, and manufactured goods arrived upriver from the west.

Devastated during World War II, Berlin was occupied by the four allied powers. The French, British, and American sectors of the city formed West Berlin, which became an island of the Federal Republic of Germany (West Germany) in the Soviet-occupied eastern half of Germany. The eastern half of Berlin was the capital of the German Democratic Republic (East Germany). During the Soviet blockade of West Berlin in 1948, the Western allies supplied West Berlin entirely by air. Tegel and Tempelhof airports, used in the Berlin Airlift, are visible in this image.

(following pages)

London

The River Thames runs through the heart of London, the capital of the United Kingdom. Because the river is tidal far upstream of London, large vessels, in the past, entered basins or docks at high tide. Watergates at the dock's entrance would be closed as the tide fell, holding enough water in the dock to keep the ships afloat. In this image, the thin dark shapes to the east of central London are the docks of London. The water in the docks appears black because it is isolated from the muddy river.

The shipping business at London's docks declined following the breakup of the British Empire and the increase in containerized shipping. Other British ports had the large area and extensive truck access that container shipping required. In 1981 the London docks were finally closed to shipping. The Docklands, as the area is called, has become a site for upscale housing and service industries. A rapid expansion of London's financial district, "The City," during the late 1980s created a demand for office space and generated a building boom to accommodate business needs.

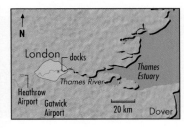

Po River Valley

Like a bolt of lightning, the Po River slashes across this image of its valley. Italy's longest river, the Po flows about 673 km (417 mi.) from the slopes of Monte Viso in the western Alps across northern Italy, and into the Adriatic Sea. A tooth-shaped delta has been formed by the Po and its distributaries. The curving sediment plume, issuing from the river's mouth, indicates the direction of currents that flow parallel to the shore.

Lake Garda, north of the river, lies in a trough carved through the foothills of the Alps by a glacier. The southern margin of the Po Valley (lower

right) is subtly revealed through road alignments, the distribution of towns and cities along the roads, and varying human impact upon the land. The differing patterns indicate land use that reflects changing terrain.

The Po Valley contains the most productive farmland in Italy and two of the country's most important industrial centers, Milan and Turin. (Neither is visible.) Venice, one of Italy's most important seaports, appears north of the delta. A causeway connects Venice to the mainland.

Venice is a provincial and regional capital, and the city is visited by more than 8 million tourists per year, attracted by the city's famous cathedrals, piazzas, bridges, palaces, and campaniles, or bell towers. Unfortunately many of the

city's architectural treasures are being severely damaged by acid rain, caused by emissions from industry, transportation, and power generation. Pigeons flutter at dawn (below right) around a square off historical Piazza San Marco.

Rotterdam

A busy, complex port, Rotterdam is the key city in the Randstad, a densely populated region of cities, suburbs, and intensive farming in the western Netherlands. Rotterdam has been an important port since medieval times because of its advantageous location near the mouths of the Rhine River and its distributaries. The Rhine and an extensive continental system of canals connect Rotterdam with industrial centers, most notably the Ruhr region of Germany, and

the markets of western and central Europe. Rotterdam has long been a transshipment point, where cargo is transferred from river barges to ocean vessels.

By the mid–19th Century the river channel that connected Rotterdam to the North Sea could not be navigated by the newer, larger steamships. A 26-km (16-mi.) channel, named the New Waterway, was dredged directly to the North Sea. The port was de-

stroyed during World War II. Most of the port and industrial facilities are a result of postwar reconstruction and expansion keyed to petroleum refining.

The most visible postwar project is the harbor at Europoort, built to handle deep-draft supertankers. Clusters of dots along the south side of the New Waterway are oil storage tanks and warehouses. Rotterdam also services offshore oil and gas production in the North Sea.

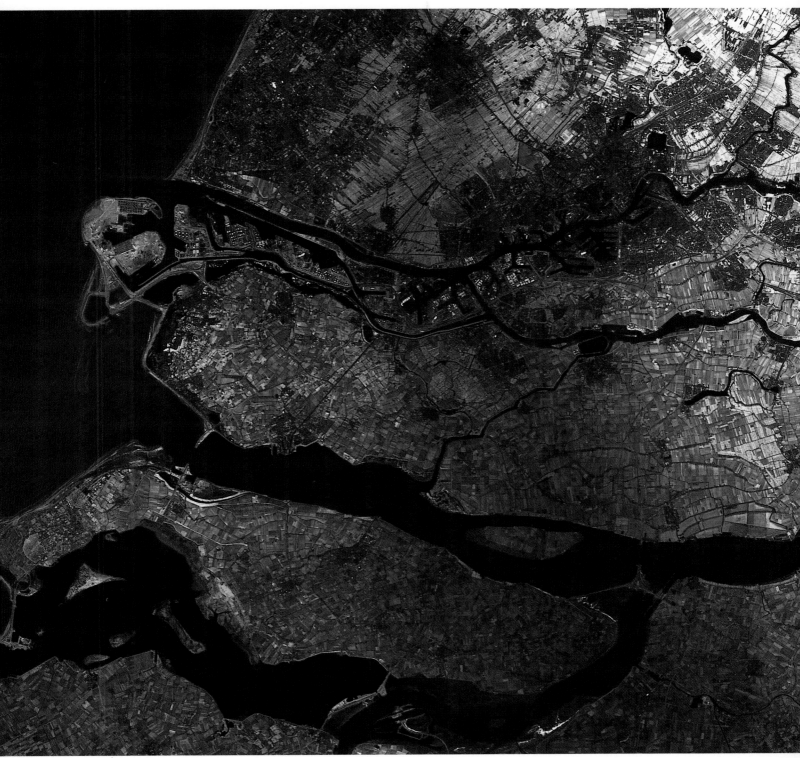

Breaking Pollution's Chokehold

INCREASED ENVIRONMENTAL ACTIVISM IN WESTERN EUROPE, especially during the last two decades, has helped to produce policies and regulations aimed at improving Europe's air, water, and soils. But the continent's dense concentration of industry still threatens the resources and quality of life in its many regions.

The problem is more acute in Eastern Europe, where the recent opening of borders and free flow of information revealed one of the most polluted places on Earth. Years of totalitarianism fostered the spread of this pollution, through pressure to produce at any cost, lack of economic incentives to conserve or operate plants cleanly, and official curbing of environmental activism. Eastern Europeans now are striving to undo the damage of several decades.

As on the other continents, throughout Europe by-products of coal-burning and automobile exhaust pollute the atmosphere and contribute to the widespread problem of acid rain. Forests in Scandinavia and West Germany have been harmed by acid rain, which also has had a corrosive effect on statues in Venice and threatens treasured buildings and sculpture across Europe. The acid washes into lakes and streams as well, where it even endangers aquatic life.

Chemical fertilizers, used in increasing quantities, also drain into rivers and lakes. (See pages 146-47.) Industrial wastes, the residue of Europe's intense commercial growth, are dumped in rivers and seas, as is sewage, often untreated.

Much of the river water in Czechoslovakia, Poland, and Romania is undrinkable, and many lakes and rivers throughout Eastern Europe are too polluted to sustain any fish. The "beautiful blue Danube" has been devastated by both Western and Eastern polluters all along its course to the Black Sea.

Like the Danube, Europe's other polluted rivers bring their tainted waters to the seas. This is especially serious in the enclosed seas, such as the Baltic, the Black, and the Mediterranean. Near

Gdansk, Poland, raw sewage and chemical wastes from refineries contaminate a coast already besieged by other problems: tourist traffic pollutes the seashore; fish are dying in greater numbers; beaches are being closed. In response, sewage treatment plants are being built and clean-up projects are under way.

Europe, like other continents, is losing forests. Woodland development in Great Britain and Scandinavia is in decline. In the Black Forest region of Germany, one-third of the trees has been damaged—partly as a result of acid rain. In some parts of Romania and Czechoslovakia the trees and other plant life are blackened with soot from power plants and factories. Half of all Czech forests have been defoliated or otherwise damaged (see opposite page).

Although given a low priority in the past, environmental issues are now leading concerns in Europe. There is a strong will and desire among its inhabitants to clean up the environment. But the costs are enormous. Pressure to produce more and more food and other vital goods continually conflicts with conservation and efforts to reduce pollution.

Poisons have no boundaries. Pollution generated by one country is often not contained within its borders. Rivers carry their loads across nations; winds hurl dust and fumes over great distances; greenhouse gases, such as carbon dioxide and methane, accumulate and influence warming of the climate on a global scale.

The crisis of environmental pollution is one that can only be solved by international cooperation. Unity of purpose must transcend narrow national interests. Perhaps Europe can provide a model for the world. If the countries of the east and the west pollute themselves, each other, and the resources they share, they risk reducing their productive and economic viability. It is hoped that the opening of Eastern Europe and the union of Western Europe will help promote standardized policies and opportunities for mutual assistance—goals that can someday, perhaps, be achieved on a global scale.

Chemopetrol chemical plant in Litvinov, Czechoslovakia, is shown at night in 1990 (above) spewing pollutants into the atmosphere, unrestricted by environmental regulations that are now being implemented.

Forest Damage

Plumes of smoke (lower right) rise from coal-fired power plants near the city of Chomutov in Czechoslovakia and sweep over the Krusne Hory (Erzgebirge) Mountains into what had been East Germany. Rectangular shapes near the smoke plumes are strip mines, the sources of the power plants' low-grade, sulfur-rich brown coal.

The haze that appears over the irregularly shaped strip mine to the north of Chomutov is from smoke. The burning of brown coal not only pollutes the air but also inflicts terrible damage on forests throughout Eastern Europe. As recently as 1978, forests of Norway spruce, which appear black in the image, covered the entire Krusne Hory region.

In the early 1980s, trees growing high on the mountains began to die. At the time of this scene (1985), there had been extensive clear-cutting of dead and dying trees in the Krusne Hory (brown and light-green areas in the mountains). Efforts to replace the lost and damaged forests have not been very successful. (T.R.W.)

Lake Balaton

Hungary's largest lake, Lake Balaton, about 90 km (56 mi.) southwest of Budapest, is ringed by numerous resorts and spas. The volcanic peninsula of Tihany extends nearly all the way across the lake to within 1.5 km (nearly a mile) of the southern shore. Several ancient extinct volcanoes lie along the lake's northern edge. These old cratered structures are now covered by vegetation, indicated by the red color on the image.

The most prominent of the volcanoes is the Badacsony, which exhibits tall columns of volcanic rock at its summit. Vineyards and farmlands are common around this volcano.

The abundance of rectangular-shaped fields reflects the concentrated farming that reigns over much of the land in the region of Lake Balaton. Due to the substantial growth of agricultural land use over the last 20 years, nitrogen and phosphorus, common constituents of fertilizers, have been washed into the lake,

contributing to an environmental problem called eutrophication. Fed by these nutrients, algae and leafy plant life clog the lake waters. When they decompose, they deplete oxygen from areas of the lake inhabited by fish and other creatures. Resort users crowd the lake shore as well, adding a burden of sewage and other pollutants to the nutrient load.

Since Lake Balaton is very shallow (only 3.4 m or 11 ft. deep), it is extremely vulnerable to rapid eutrophication. Many programs and research projects have been undertaken in Hungary over the past two decades to study this problem. Although much is still to be done, the water quality of the lake has been improved over the last few years through sewage treatment projects and control of agricultural run-off.

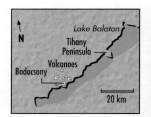

Jet Contrails

Jet contrails blanket the skies over Germany, Switzerland, Austria, Czechoslovakia, and northern Italy. The occurrence of so many long-lived contrails at one time is somewhat rare and reflects specific meteorological conditions. But the image demonstrates a phenomenon stemming from the growth of the worldwide commercial air transportation industry.

Contrails form in cold moist air when water vapor from aircraft engines condenses. The contrails in the image trace some of the most heavily trafficked air corridors of Europe. Trails turning sharply in the center of the image indicate routes toward Amsterdam and Frankfurt, Europe's third-busiest airport (after London and Paris). Diagonal trails in the south are probably heading for Paris. The image was collected when Germany was still divided and East-West relations were cool. The contrail-free band between Eastern and Western Europe, although possibly due to weather conditions, may reflect the lack of free movement and communication between East and West.

What are the environmental effects of jet contrails? Scientists in Germany are studying satellite imagery to get an answer. They hope to learn if contrails have an impact on local or global climate by increasing cloud cover.

Chernobyl

The bright red square (above, center) shows the intense heat emanating from the core of the Chernobyl nuclear power plant three days after the explosion that stunned the world. In the early days of the Chernobyl crisis, the accident was observed in satellite imagery. This SPOT image is a combination of information in visible and thermal wavelengths.

The town of Chernobyl in Ukraine lies along the Pripyat River about 130 km (80 mi.) north of Kiev and about 700 km (430 mi.) from Moscow. In the early morning hours of April 26, 1986, one of the plant's four reactors exploded, releasing large amounts of radioactive materials into the atmosphere. The 2,000-ton

reactor lid was dislodged and material from the core was blown out. Courageous workers died of radiation exposure shortly after putting out a fire on the roof. But the core continued to burn and glow until May 6. In about a week the reactor (below) cooled down enough for the radioactive emissions to be brought under control. Because these materials were released over such a long period at high altitudes, they reached as far as the United States, Canada, and Japan (in relatively small quantities).

Approximately 100,000 people were evacuated from areas around the site on the second day of the crisis. Clean-up of the region has proved very difficult. Windblown dust

and soil spread the contamination. Tar roofs in and around Chernobyl collected dangerous levels of material, while rains carried radioactivity into drinking water and deep into farmland soils.

After the explosion, about 30 plant employees and emergency workers died of direct radiation exposure. Many more people will likely die from delayed cancers over the next

decades. There may have been an increase in leukemia and thyroid cancer in children, and doctors fear long-term increases in birth defects and other diseases.

Soviet scientists have explored the stark ruins of the reactor itself in order to account for all of the highly radioactive core material and to verify that another chain reaction is unlikely.

FROM SEA TO SEA

THREE OCEAN BASINS—THE ARCTIC, ATLANTIC, AND PACIFIC—SURROUND North America. Bounded by water, the bountiful continent of North America is like a great island that developed in isolation. Europeans applied the term New World to the continents west of the Atlantic. Though hardly new to the inhabitants who greeted the Europeans, the lands were indeed new to people who first arrived from Asia.

Canada and the United States occupy most of the North American landmass north of Mexico. Two European nations, however, have a presence in North America that dates back to colonial times. Greenland is a self-governing province of Denmark, with an economy based on commercial fishing and some traditional hunting. St. Pierre and Miquelon, two small islands south of Canada's Newfoundland coast, form an overseas territory of France. The islands have historically been a base for French commercial fishing in the North Atlantic.

The northern limit of North America, the Earth's third largest landmass, is a small island off the north coast of Greenland. Although that dot of land is closer to the geographic north pole than any other land on Earth, the southern boundary of North America is not that easily defined. Mexico lies on the North American plate and is physically part of the continent's landmass. But cultural geographers often include Mexico with Central America and Panama in a region known as Middle America. And that is where Mexico is presented in this book. (See Middle America chapter, page 184.)

On the western boundary, Alaska's Seward Peninsula almost divides the Arctic and Pacific oceans as it projects toward Siberia. The Bering Strait, at its narrowest point, separates North America from Asia by a distance of only 85 km (53 mi.). The shallow strait is approximately 50 m (160 ft.) deep and often choked with ice. Here began the peopling of the Americas.

About 2 million years ago the Earth's climate cooled and the so-called Ice Age began. During this epoch (more properly known as the Pleistocene) great ice sheets advanced and retreated over the northern continents. A significant amount of the Earth's water was frozen in the ice sheets as they spread. At times sea level dropped as much as 100 m (330 ft.) below its current level. This fall in sea level opened land bridges between some of the continents. Alaska and Siberia were connected, making possible one of the most important human migrations of prehistoric times.

Possibly as early as 60,000 years ago hunter-gatherers crossed the Bering land bridge from Asia to North America. By the end of the Pleistocene, about 10,000 to 15,000 years ago, the last great continental ice sheets had retreated. Sea level

NOAA, AVHRR MOSAIC,

TERRA-MAR RESOURCE INFORMATION

SERVICES, INC.

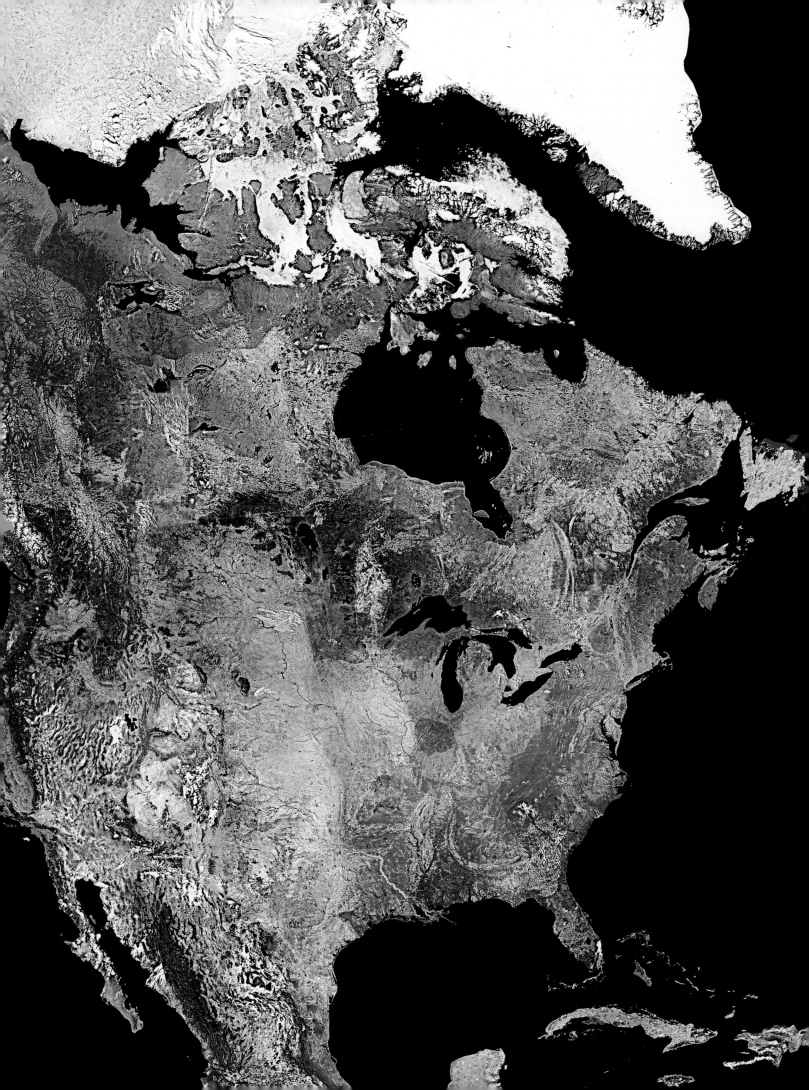

rose and, relieved of the tremendous mass of ice, the continents sprang back. North America rebounded about 300 m (980 ft.) along the eastern shore of Hudson Bay, where the continental ice sheet was probably thickest. The Bering land bridge was submerged, and Asia and North America again were separated by a strait. By then, however, the first discoverers of the New World had spread throughout the continent. As the renowned explorer Thor Heyerdahl once said, the first person to die on North American soil was Asian-born.

Descendants of the Bering Strait migrants, ancestors of American Indians, had the continent to themselves for millennia. Debates go on endlessly about when and which Europeans, Asians, or Pacific Islanders were first to reach the continent after the original discoverers. What is agreed upon follows:

Greenland is known to have been explored by Norsemen in the 10th Century. In 982 AD, Eric the Red journeyed to the island after being exiled from Iceland for several years. In 985 he brought back to his homeland news of the land he called Greenland. (By some accounts, he named it Greenland to encourage settlement; other accounts say he named it for the vegetated fjord valleys.) Soon he traveled back to the island with 25 ships carrying people and supplies. The settlers flourished in the beginning, but through the centuries, climate change and trade problems caused the colonies to decline. By the late 15th Century, communication with the settlements ceased, their fate unknown.

Evidence from an archaeological site in Newfoundland proves that Norse colonists had settled in North America by about 1000 AD. The Norse colony at l'Anse-Aux-Meadow was short-lived and its impact on North America was not profound.

Not until the 1492 voyage of Christopher Columbus did the New World begin to suffer the consequences of contact with Europe. In the first half of the 16th Century, just a few years after Columbus died, the Spanish conquered Mexico, Central America, and part of South America in their quest for gold and silver. Many thousands of Indians died fighting the conquistadors. During and soon after the conquest, possibly as many as 10 million Indians died from new diseases that were carried to the Americas by Europeans and Africans.

North America north of Mexico was initially spared outright conquest. This area of the continent attracted little attention after early expeditions, probing northward, failed to encounter materially wealthy cultures, such as the Aztec and Inca empires that had been found and plundered to the south. In the 17th Century Europeans, and the Africans they enslaved, began arriving in number on the eastern coast of North America. During the next two centuries, westward expansion of these newcomers changed much of the North American landscape. People from throughout the world settled North America, creating nations that have had a global influence on technological and commercial development.

Seen from space (previous page), North America by day looks as untouched as it did before people began arriving. But at night (see pages 174-75) city lights produce a shining portrait revealing the continent's urbanization. While the Industrial Revolution demanded great new cities as centers of industry, manufacturing, and commerce, the post-industrial age has decentralized and spread urban growth all over the land.

ARCTIC OCEAN

GREENLAND

BAFFIN BAY

p.155 Greenland

LABRADOR SEA

p.170-71 Lake Laberge

HUDSON BAY

p.170 Northlands

NORTH AMERICA

CANADA

p.156-57 Manicouagan Crater

p.174-75 City Lights

p.182-83 Vancouver

p.177 Montreal

p.156 Gaspé Peninsula

St. Lawrence R.

p.168-69 Mt. St. Helens

p.163 Lake Michigan

p.159 Appalachian Ridge

p.162-63 Great Salt Lake

p.158-59 Cape Cod

p.173 New York City

p.180-81 San Francisco

Missouri River

Colorado R.

UNITED STATES

p.176 St. Louis

Ohio River

p.178-79 Dulles International Airport

p.164-66 Grand Canyon

p.166-67 Imperial Valley

Mississippi R.

ATLANTIC OCEAN

p.180 Houston

p.161 Mississippi Delta

p.160 Florida Everglades

Rio Grande

GULF OF CALIFORNIA

MEXICO

GULF OF MEXICO

CUBA

THE TERRA... mountains to... swamps and... four physiog... lands of the... chains that b...

The Cana... eroded rock... extends in a... in the Adiro... found in the... ever identific...

The vast... lowlands lie... the Canadian... continent's... ranges and t... the east. In th... ciers scraped... face, leaving... layers of ma... great distan... the Mississip... continue to... The Great P... includes that... plains as w... terrain, stre... Grande in th... of western C...

Mountain... of the centra... on the west... tent. They a... ing the Roc... and the Sier... two segment... are divides... valleys. The... in the west... marks the... the cordille... Columbia P... the Colorad...

Greenland

Long fjords, partially filled by glaciers, line the coast in this Space Shuttle photograph of southeast Greenland, taken in March, before the spring thaw. Pack ice floats off the coast and "fast ice" is attached to the shore. The open water at the mouths of fjords forms due to winds blowing down the valleys. As ice breaks off at the ends of the fjords, it is blown eastward until it is blocked by the barrier of ice along the coast.

Most of Greenland lies north of the Arctic Circle, and it is the second-largest island in the world (after Australia). A massive ice sheet, second in size only to that of Antarctica, covers about 80 percent of the island's area.

Glaciers flowing down from the ice sheet fill the narrow fjords. Large segments calve off the ends of the glaciers and float out into the ocean, making Greenland the major source of North Atlantic icebergs that endanger shipping.

glacier — fast ice — N

glacier — pack ice

GREENLAND — 20 km

Gaspé Peninsula

In this Space Shuttle photograph, snow highlights cleared land surrounded by darker patches of forest on the Gaspé Peninsula, the most sparsely populated region in southern Canada. The peninsula's name is probably derived from the Micmac term gespeg, which means "end of the world."

The Gaspé's thin soils and severe winters have kept settlers away since Jacques

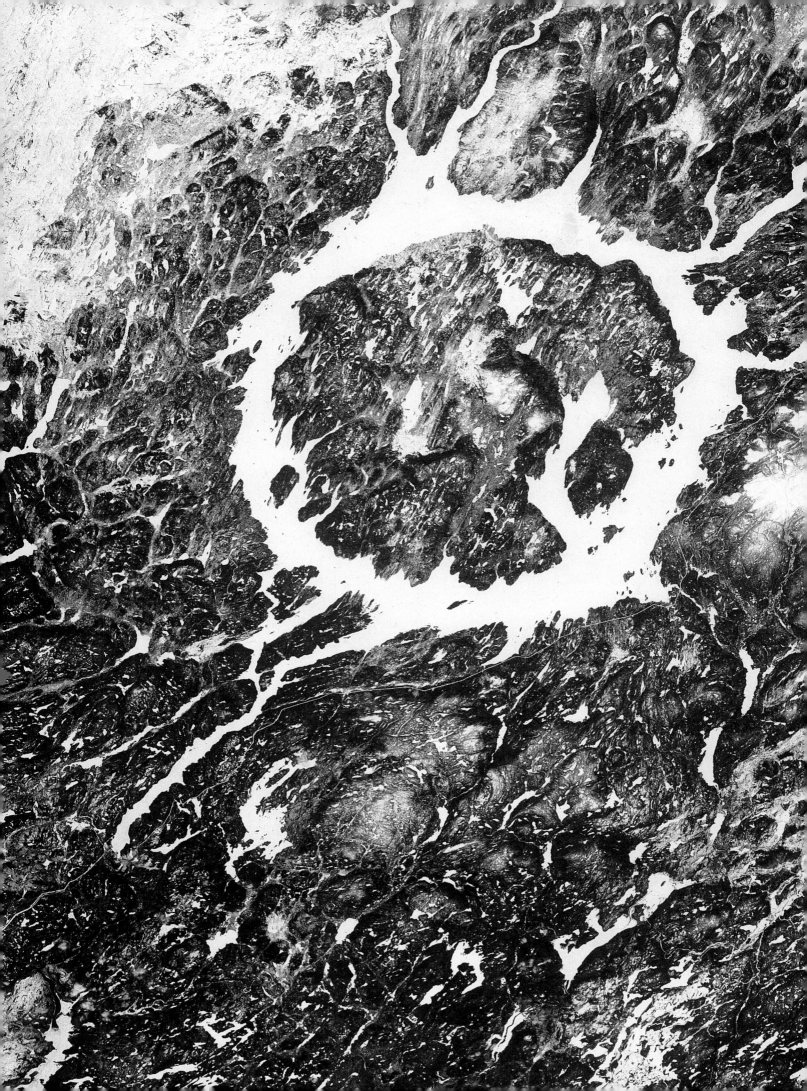

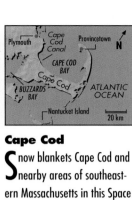

Cape Cod

Snow blankets Cape Cod and nearby areas of southeastern Massachusetts in this Space Shuttle photograph. This area, like most of Massachusetts and the northeast, has figured prominently throughout American history. In 1620, the Pilgrims landed first at the site of Provincetown and then settled at Plymouth. Part of Nantucket Island, the whaling capital of the world during the 18th and 19th centuries, appears at the lower left.

Cape Cod was formed by materials transported and deposited at the forward edges of Pleistocene glaciers. More recently, waves and currents shaped portions of Cape Cod's shores.

The Cape Cod Canal, visible as a wavy dark line, connects Cape Cod Bay to Buzzards Bay. The canal was completed in 1914 by the Army Corps of Engineers. Pilgrim Miles Standish first suggested such a canal to avoid the treacherous voyage around the cape.

Cape Cod has become a haven for vacationers who must cross the canal to spend time at beaches or resorts. Some 40,000 acres on the eastern portion of Cape Cod have been set aside as a national seashore. (R.A.C.)

Appalachian Ridge

This radar view reveals the distinctive structure of the Appalachian Mountains near Harrisburg, Pennsylvania. This Ridge and Valley Province, as geologists call the region, is characterized by bands of mountains formed by the folding of layers of sedimentary rocks. As the terrain eroded, more resistant layers formed ridges, while more easily weathered rocks were carved into valleys. Rivers cutting across the ridges have provided the easiest routes of travel through the uneven terrain. Included in the sweep of the Appalachians are the White Mountains of New Hampshire, the Green Mountains of Vermont, the Berkshires of New York and Massachusetts, the Catskills, the Alleghenies, the Blue Ridge Mountains, and the Great Smokies.

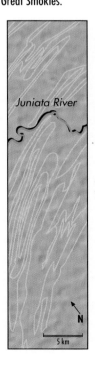

Florida Everglades

Lake Okeechobee looks like a hole in southern Florida in this portrait of land and sea. Miami lies to the south. Cultivated fields cluster near the lake, which covers about 1872 sq. km (720 sq. mi.) and fills a shallow basin to an average depth of less than 6 m (20 ft.). Lake Okeechobee is a natural

the water that flows into the Everglades, which is now dry for much of the year.

The Everglades National Park was established at the southern and southwestern end of the region. The area supports alligators, panthers, dolphins, manatee, bald eagles, and such water birds as egrets, herons, ibis, pelicans, and spoonbills—all dependent on the ever-diminishing flow of water through the Everglades.

Mississippi Delta

Turbid waters surround the "bird's foot" form of the Mississippi River Delta, where river channels empty into the relatively deep water of the Gulf of Mexico. Differing densities of suspended sediments produce the varied colors. Near-shore currents carry away the sediment-laden water, producing clouded patterns.

The delta grows through extension of the major distributaries (channels that branch away from the primary channel). Dramatic expansion of a river delta occurs when the sedimentation rate is much greater than the destructive action of waves and tidal currents.

The Mississippi River has been flowing into the Gulf of Mexico for millions of years. But the delicate delta landforms do not survive for long when the major distributaries shift position. On either side of the modern Mississippi Delta are remnants of older deltas, each once the center of delta growth. They date from the rise in sea level that followed the last ice age. The Chandeleur Islands, which appear as a crescent to the right of the image, are formed by waves reworking the upper part of a delta that was active from 600 to 4,700 years ago.

In the early 1950s the river was diverting an increasing amount of its waters to the Atchafalaya River, which branches west from the Mississippi at a site more than 217 km (130 mi.) upriver from New Orleans. This change indicated that the river was once again preparing to shift its main site of delta growth. In 1955 the Army Corps of Engineers began a gigantic project to prevent major diversion into the Atchafalaya. Continual construction is needed to keep the Mississippi River flowing through the major port of New Orleans to its present delta. To avoid shoals near the river's mouth, which prevent large freighters and tankers from passing upstream to the port, the 123-km (76-mi.) Mississippi River–Gulf Outlet was built in the late 1960s. Parallel to the river's east bank, it connects New Orleans to the Gulf for deep-draft vessels. (J.R.Z.)

Great Salt Lake

A railroad causeway through the middle of the Great Salt Lake dramatically draws a line visible from space. The variation in color is caused by differences in the concentration of salt in the two halves of the lake. The causeway interrupts the mingling of the lake's northern and southern waters. The southern half of the lake receives more run-off water from neighboring mountains.

The largest inland body of salt water in the Western Hemisphere, the Great Salt Lake, formed as a much larger lake spreading into what became Utah, Idaho, and Nevada. When it dried up, following the ice ages, salt flats formed as soils, rich in sodium and magnesium, on the former lake floor.

Industries now enhance the natural processes of salt formation. On the eastern and southwestern edges of the lake are desiccation ponds, large evaporite pans in which lake water is let in and allowed to evaporate, leaving layers of salt that is later refined.

At the northeast shore of the lake is Promontory Point, the site of the completion of the first transcontinental railroad, where the Union Pacific and Central Pacific met in 1869. The causeway that divides the lake, completed in 1959, connects the towns of Ogden and Lucin.

In 1847, Mormon settlers first entered the valley through a pass in the Wasatch Mountains and began the settlement now known as Salt Lake City, visible at the bottom center of the image. (T.A.M.)

Lake Michigan

A heat-measuring instrument produced this springtime image of Lake Michigan (below), the third largest of the Great Lakes and sixth largest lake in the world. Curving around the west coast of the lower peninsula of Michigan, it joins Lake Huron through the Straits of Mackinac, visible in the upper right.

Light tones represent warmer areas; darker tones depict cooler regions. Lake waters are warmed by absorption of heat as seasonal temperatures rise, and by warmer waters coming from tributaries and industries along the shore. The warm zone near the shore tends to expand toward the deep cold central region through the summer. Warming in Green Bay and Lake Winnebago to the west may be abetted by a lack of circulation. The heat imaging shows the relatively warm waters of the Mississippi River and the urban heat islands generated by Chicago and Milwaukee.

Concentration of population and industry around the Great Lakes has brought with it serious and long-ranging environmental problems because the lakes have been dumping grounds for toxic wastes for decades. Even though efforts by the United States and Canada to clean up the lakes have improved water quality and brought life back to areas once thought dead, the lakes remain seriously threatened.

There is still great concern about the long-term health hazards that may result from eating some species of lake fish. The pollution will not be rectified overnight.

(previous pages)

Grand Canyon

The layered walls of the Grand Canyon and the stone-carving Colorado River etch the Arizona landscape in this Space Shuttle photograph. Faint plumes from a small forest fire smudge a shoulder of the north rim (lower center). Dark areas represent the change from grasslands to pine forests. Patches of clearcut Kaibab National Forest can be seen near the center area of the south rim. Faults cutting across the canyon are visible as linear changes in color or as elongated drainage basins.

A person walks through time hiking down one of the trails leading into the canyon, which is 900 to 1800 m (3,000 to 6,000 ft.) deep in places. The most popular trail, the Bright Angel, begins in lime-stone rock, which is older than 225 million years. Deeper into the canyon, layers of other rock are even older, and fossils become simpler and more primitive. Eventually, at the bottom, the rocks are older than 570 million years and come from a time when only single-celled organisms lived in the oceans. [R.A.S.]

Mount St. Helens

The 160 km per hour (100 mph) landslide and a volcanic blast traveling at 320 km per hour (200 mph), combined to level and sweep away 600 sq. km (230 sq. mi.) of forest. (J.R.2)

Northlands

The northernmost reaches of North America are visible in this mosaic of images that also covers the North Pole and northern Europe and Asia. Visible North American lands shown include the snow- and ice-covered island of Greenland, much of Canada, and Alaska. Southampton and Coats islands lie on the northern edge of Hudson Bay. The bay is frequently fog-covered, which may be the reason it is partially obscured in the image.

Farther west can be seen Great Bear Lake, whose northern arms lie on the Arctic Circle. One of the largest lakes in Canada, it is named for the bears who live and breed along its shores and feed on the abundant fish.

Many of these lands lie above the Arctic Circle, a line of latitude located at roughly 66°30'N. Because of the Earth's tilt on its axis, points above this latitude experience

at least one day a year when the sun does not rise or set. The farther north one travels, the longer the periods of darkness or light persist.

Near the continent's west coast, snow-covered mountains snake across the terrain. Canada's Mackenzie Mountains extend up from the Rockies. Alaska's Brooks Range, lying above the timberline, has no forests. North of the range along Alaska's coast is Prudhoe Bay, the site of vast oil reserves. From here the trans-Alaskan pipeline is routed through a pass in the Brooks Range and on to Valdez in the south. On the western edge of

the North American continent can be seen the Bering Strait. The narrow passage is generally thought to be the site of a land bridge over which Asian ancestors of Native Americans crossed to populate the North American continent.

Above, Ellesmere Island is the northernmost point in Canada. Bordering the northern coasts of North America, Europe, and Asia is the Arctic Ocean. Small as oceans go, it is almost completely surrounded by land. The geographic North Pole lies near the center of the image in an area of drifting ice. For six months of the year the Pole is cloaked in continous darkness, and for the other six the sun does not set.

Lake Laberge

Long and narrow, Lake Laberge fills a steep-sided valley through which the Yukon River flows in the snow-capped mountains of the southern Yukon Territory. The Yukon River flows northward across the Yukon Plateau and then turns westward into Alaska, where it crosses the state and empties into the Bering Sea.

In 1896 gold was discovered along the Klondike River, a tributary of the Yukon that joins the river about 430 km (266 mi.) downstream of Lake Laberge. Thousands of prospectors poured into the wild Yukon Territory along two main routes to the fabled gold of the Klondike: up the Yukon River by paddle steamboat from Alaska or across the Coast Mountains from Alaska's panhandle and down the Yukon River by raft or paddle steamer.

Prospectors on the down-

stream route traveled the length of Lake Laberge, using sleds and dogs on risky winter trips when death from immersion in the frigid water was a constant danger. The Yukon's inlet and outlet at Lake Laberge were particularly treacherous because the river's currents mix with a layer of relatively warm lake water. This mixing causes the ice at the inlet and outlet to be much thinner than ice in the middle of the lake.

To hurry the spring breakup for the river passage, riverboat crews would spread ashes on the ice. Instead of reflecting solar energy, the ice would melt as the dark ashes absorbed the sun's radiation and warmed the ice beneath.

Although "Klondike" will always suggest the rush to riches, the gold boom ended by 1910, and riverboats stopped running on the Canadian reaches of the Yukon in the 1950s.

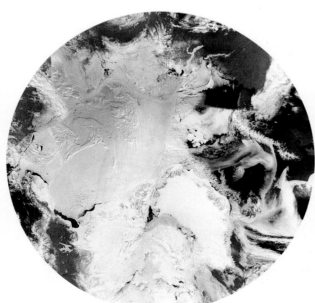

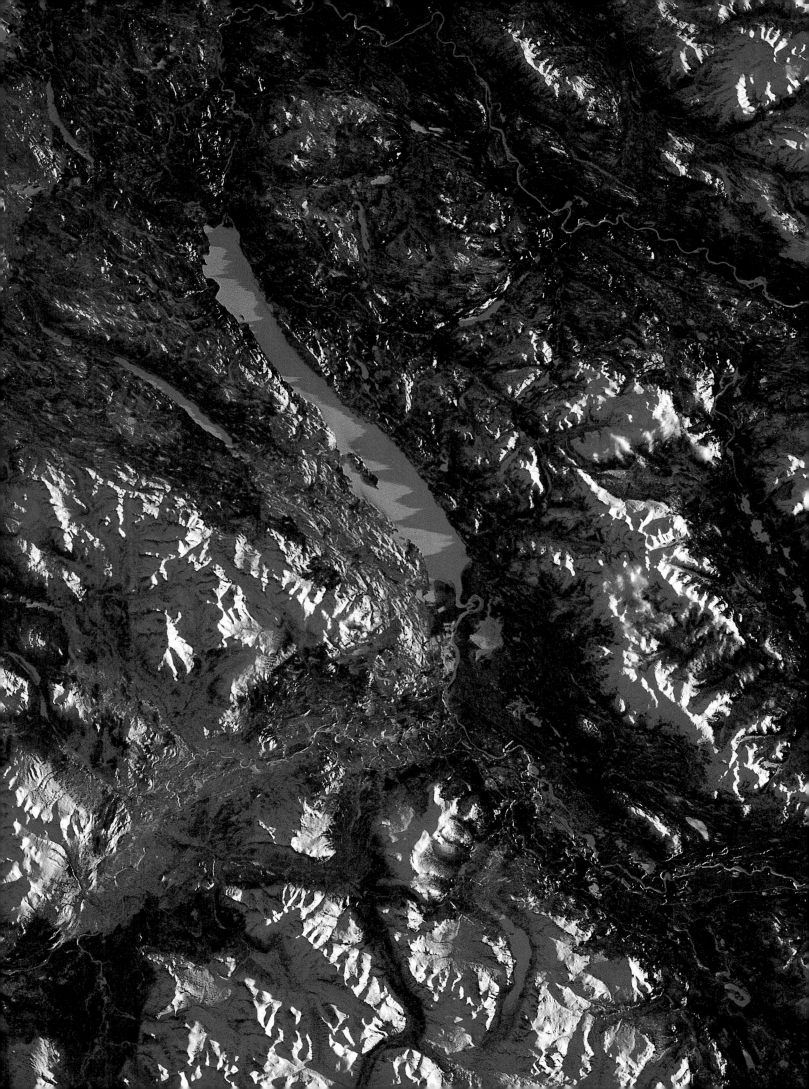

URBANIZATION IN NORTH AMERICA

IN 1910 THE U.S. CENSUS FIRST USED THE TERM "METROPOLITAN district" to define a continuous settlement that included both city and suburbs. In 1964 a new term, megalopolis ("great city"), was used to describe some parts of North America, such as the northeastern United States, where metropolitan areas were coalescing into a nearly continuous urban mass. By the late 1980s about 76 percent of the population of Canada and the United States lived in metropolitan areas.

In the mid–18th Century, most of North America's cities were ports on the East Coast. New York, Boston, Philadelphia, and Charleston served primarily as gateways for the flow of raw materials from North America and finished products from Europe.

Following the American Revolution, inland mill towns were developed by manufacturers dependent upon water power. The towns were along the boundary of the Atlantic coastal plain and the piedmont ("foot of the mountain") plateau. This boundary is known as the fall line because many of the continent's eastern rivers reached tidewater after descending a final set of falls or rapids. Ships could ascend rivers up to the fall line, so the rivers not only generated energy, but also connected the mill towns to the port cities.

Balanced more than 303 m (1,000 ft.) above Chicago, a technician repairs an antenna cable on Hancock Center's "Big John" building, one of many boldly designed skyscrapers that characterize the city.

Canals were built along the few feasible routes through the Appalachians. The Erie Canal connected the Great Lakes with the Hudson Valley, making water transport possible between the Midwest and New York City. West of the Appalachians, the Mississippi River became the trunk of a transportation network that included the Missouri, the Ohio, and other rivers. This network connected the coastal port city of New Orleans with such inland ports as Memphis, St. Louis, and Cincinnati.

The Industrial Revolution was the next major catalyst to urban growth in North America, and it was a revolution that radically altered the continent's urban landscapes. The energy demands of the pre-industrial cities were modest. Firewood was used for heating and cooking. Iron production was limited to small blast furnaces at the site of ore deposits. The forests surrounding the blast furnaces were cut to produce charcoal for fuel and were a source of carbon for pig-iron production.

The Industrial Revolution had four major needs: water, coal, iron, and labor. North America already had an abundance of the natural resources. The demand for labor was met by the waves of immigrants who arrived from Europe and Asia in the 19th Century. A growing railroad system made it economical to transport bulky raw materials overland. Water transport, more efficient with the use of steam propulsion and larger vessels, moved raw materials cheaply to the factory, mill, or blast furnace.

When steam replaced water as industry's prime source of energy, the urban landscape soon included the smokestacks and water towers of large factories. Worker housing, often of uniform construction, was usually built within walking distance of the factory. By 1920 a belt of industrial cities extended from southern New England west to Chicago, including southern Ontario. In 1920 the urban proportion of the total population was approximately 51 percent, compared to about 6 percent in 1800.

Oil eventually replaced coal as an industrial power source and became fuel for the automobile, which revolutionized transportation. New roads were built. In western and southern regions outside of the historical industrial core, younger cities more easily accommodated the automobile.

North America prospered in the 1950s, and automobile ownership became common. Because urban workers were able to commute by automobile from the suburbs, many urban areas lost population. Industry also fled city confines for growing room and easy access to superhighways. Today the term "inner-city" suggests a run-down urban core populated by the poor.

New York City

Manhattan Island, in the center of this image of New York City, is bordered on the west and east by the Hudson and East rivers, which extend northward from the ship-filled New York Bay. To the north, the Harlem River separates the island from the Bronx. Standing out clearly in contrast to its densely developed surroundings is the red rectangle of Central Park. Also visible are the George Washington Bridge, crossing the Hudson River into New Jersey, and several other bridges connecting Manhattan to Brooklyn and Queens.

The famed Brooklyn Bridge crosses into Brooklyn Heights, the site of the first Dutch settlement on Long Island in 1636. At the southern end of Brooklyn is the amusement park, Coney Island. A glance at the image, however, shows that the area is not an island. When the early Dutch settlers called it Konijn (Rabbit) Eiland, it was indeed surrounded by water. Gradually, however, the creek that separated it from Long Island clogged up with silt and mud, and the two islands were joined. Southeast of Coney Island, Rockaway Inlet leads to Jamaica Bay. In the northeast, the East River joins Long Island Sound at Throgs Neck, site of the most easterly bridge between New York's mainland and Long Island.

City Lights

In a mosaic of nighttime images, North American cities glitter from coast to coast. This imagery was collected at night by Defense Meteorological Satellites. Major cities show up as bright regions illuminated by lights shining throughout densely industrialized and urbanized regions.

The most highly concentrated megalopolis seen here is in the northeast region, stretching from Boston through New York (below) and Philadelphia to Baltimore and Washington, D.C. As the cities have grown and the areas outside the city limits have become increasingly built up, densely developed urban networks have expanded and overlapped.

To the west, around the area of the Great Lakes, are other highly concentrated urban centers. This region, rich in resources, has shipping access to markets along the lakes and overseas. Growth has merged the suburbs of Chicago and Milwaukee on the western side of Lake Michigan. On the northwest shore of Lake Erie is the vast complex growing up around Detroit. On the western tip of Lake Ontario is Toronto, and along the St. Lawrence River, on its way to the sea, are the cities of Montreal and Quebec.

The density of bright urban islands is lower throughout the continental interior. This is the less heavily populated region of the Great Plains of the United States and Canada.

The coasts are outlined by urban centers, which literally reflect the traditional commercial growth of port cities.

The west coast is highlighted in the north by Vancouver, Seattle, and Portland. Farther south, San Francisco and Los Angeles stand out. In the southern United States, Houston and New Orleans are clearly visible. On the Florida coast, a narrow strip marks Miami and other nearby communities that follow the line of popular beaches northward to Palm Beach for more than 120 km (75 mi.).

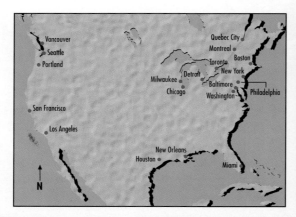

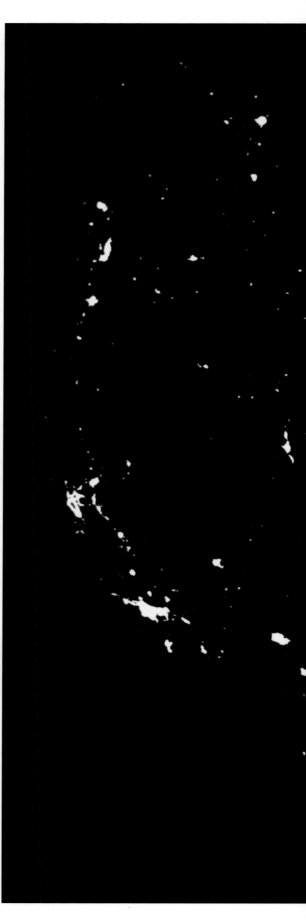

St. Louis

In a grassy park directly on the riverfront (red color) rises the Gateway Arch (above), symbol of the city that was the gateway to the western United States for fur traders and explorers. The Missouri Botanical Gardens can be seen in the southern part of the city, and Forest Park shows clearly as the large red-toned area west of the city center. The site of the 1904 Louisiana Purchase Exposition, Forest Park now is the home of the St. Louis Zoo and many other recreational and cultural facilities. To the northwest, in the tree-covered suburbs of St. Louis County, Lambert–St. Louis International Airport shows clearly.

Established as a French fur-trading post in the 18th Century, the city was also the starting and ending point of the Lewis and Clark expedition. St. Louis became a major river port for settlers and goods headed west after the War of 1812. Extensive industrial expansion followed the Civil War. The old city, named after Louis IX of France, has retained some of the architecture as well as many place names and traditions of its French cultural heritage.

Broad agricultural lands surround St. Louis. River bluffs, oxbow lakes, and the conjoined channels of the Illinois, Missouri, and Mississippi rivers mark the great floodplains of these major waterways. Patchworks of fields (light blue) fringe the rivers, especially in the wedge of land between the Mississippi and Missouri. Like all floodplains, those found in and around St. Louis are richly fertile. They also define the place where flood waters must go when the rivers run high in the spring. St. Louis, like all river cities, has learned to co-exist with the sometimes capricious waters that bring agricultural and commercial wealth. (P.A.J.)

Montreal

History, in the form of long riverfront lots, is engraved in this image of the St. Lawrence Valley (below). In 1608 Samuel de Champlain established a French colony here, later known as New France. Several large rivers flow into the St. Lawrence. Early settlers cultivated the fertile soil of the lowlands in long, thin lots.

This distinctive system of land division is visible around Montreal and along the St. Lawrence Valley. The lots were granted to the settlers in strips, each with a narrow frontage on a river. Farmers needed access to a river because they used boats to bring the crops to market, and the French quickly adopted the indigenous canoe as the most efficient craft. (Similar lots were also granted along the bayous of Louisiana near the French colonial city of New Orleans.)

Montreal was founded as a mission and trading post in 1642 by the Sieur de Maisonneuve. The city is located on an island, the Île de Montréal, in the St. Lawrence River and takes its name from a hill that overlooks the river. Because rapids in the St. Lawrence near the Île de Montréal stopped sea-going vessels from sailing upriver, Montreal became a transshipment point for goods coming down river from the Great Lakes and for cargo arriving from overseas ports.

In the 1950s the St. Lawrence Seaway was completed, and it became possible for ships to enter the Great Lakes from the lower St. Lawrence River. The seaway passes Montreal along the opposite bank of the river.

Montreal is the largest city in Canada and is a major North American commercial, industrial, and cultural center. About two-thirds of Montreal's 1 million residents speak French as their first language, making it the second-largest French-speaking city in the world. Beginning with the Quebec Act of 1774, legislation has helped French Canadians maintain their cultural identity for two centuries. Since the 1960s a vocal Francophone separatist movement, to guarantee the endurance of Québecois culture, has argued for Quebec's independence from Canada.

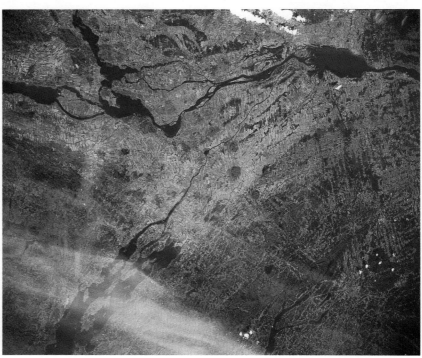

Dulles Airport

Urban growth produces some of the most dramatic changes to the Earth's surface that are visible from space. These images show a 900-sq.-km (350-sq.-mi.) area around Dulles International Airport, located about 40 km (25 mi.) west of Washington, D.C., in Virginia. Dulles opened in 1962 and began by handling mostly long-duration domestic and international flights. Through the years traffic at the airport has grown, and the surrounding area has been increasingly developed. By 1989, 10.8 million people traveled through Dulles yearly.

The images are Landsat scenes acquired in November 1982 (below) and October 1989 (opposite page). Urban areas (in shades of blue) contrast sharply with forests (in green, brown, and black) and grassland and cultivated land

(in light red and white). Comparison of the two images reveals a striking increase in urban development along many of the highways east, north, and south of the airport. The most remarkable increase in urban development is along Route 28, a main highway running generally north-south through the center of the scenes. Urban growth has been accompanied by the expansion or widening of most of the major highways in the area. Particularly prominent is the expansion of the Dulles Access Road to the east of the airport.

The amount of change in the Dulles area can be estimated by generating land-cover classification maps using the differing image data for the two years. Such an analysis indicates that urban areas increased by approximately 9 percent during the seven years between the two scenes. The development of this region over that period had a significant impact on the area's forests. About 8 percent of the forest that was present in 1982 has been lost to urban growth. (T.R.W.)

Houston

The modern urban sprawl that marks Houston is illustrated in this photograph taken from the Space Shuttle (below). Suburban developments that surround the city are linked by a beltway that is more rectangular than those that encircle many other North American cities. South of the city lies the National Aeronautics and Space Administration's L. B. Johnson Space Center, responsible for all

aspects of manned spaceflight from launch to landing.

Galveston, the hurricane-battered island city that was once home to pirate Jean Lafitte, lies 80 km (50 mi.) south of Houston. In September 1900 a hurricane unannounced by weather forecasters killed 12,000 people in the area. Galveston Island protects the inland waterway and Galveston Bay from storm winds, providing a safe harbor for ships using the channel from Galveston to Houston. The Houston Ship Channel, completed in 1914, provides a path through Buffalo Bayou and Galveston Bay for the importing and exporting of manufactured goods as well as for the shipment of petroleum products, a

major industry of the region.

Petrochemical plants, oil storage tanks, mills, and factories line the channel. Industrial wastes and treated sewage pollute the channel and threaten Galveston Bay. (T.A.M.)

San Francisco

Two areas of limited urban development, Golden Gate Park and the Presidio, contrast sharply with the surrounding developed sections in this view of the San Francisco Bay area (opposite page). Ponds with distinctive green and red colorations are areas where salt is obtained through evaporation of the brackish bay waters. Several groups of "mothballed" World War II cargo ships are moored side by

side along the northern shore of Suisun Bay, visible near the mouths of the Sacramento and San Joaquin rivers.

The narrow entrance to the bay provides an excellent natural harbor, one of the busiest on the western coast of North America. But the topography that forms the bay is a manifestation of geologic forces that threaten the city. The mountains owe their orientation to the San Andreas and Hayward faults, located on opposite sides of the bay. The San Andreas fault (above) is the junction between the North American and North Pacific plates of the Earth's crust, with the North Pacific side moving an average of 1 cm per year to the north. Consequently, Point Reyes is gradually moving up the California coast and may eventually become an island separated from the continent.

Movement along the faults in this area often occurs as sudden shifts. The major earthquakes of 1906 and 1989 resulted from displacements along the San Andreas fault, and the Hayward fault poses an additional threat. The 1989 earthquake demonstrated that damage can be caused not only by proximity to the source of the quake but also by the kind of ground that the tremors move through: the greatest structural damage in the San Francisco area occurred on landfill used to reclaim portions of the bay. (J.R.Z.)

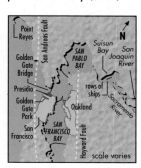

Vancouver

Sediments swirl in the busy harbor of Vancouver, Canada's third-largest city. Just north of the U.S.-Canadian border, Vancouver lies between two waterways, Burrard Inlet to the north and the Fraser River to the south. Additional port facilities are at nearby Roberts Bank, where huge piers, built to handle large commercial vessels, jut out into the sea. Together, these two harbor areas comprise the busiest port on the west coast of North America.

Also visible in the image is the city of Victoria on the southeast coast of Vancouver Island. Off Victoria, looking like pieces of a scattered jigsaw puzzle, are the San Juan Islands. Southeast lies Whidbey Island. A visible highway links two Washington coastal cities, Bellingham and Everett. At the bottom of the image, across the Strait of Juan de Fuca from Victoria, is part of Washington's Olympic Peninsula.

Victoria was first established in the 1840s as a fur-trading center by the Hudson's Bay Company. The city later became the capital of the province of British Columbia. Today a busy ferry system transports heavy traffic between the cities of Victoria and Vancouver.

Vancouver (below) began as a town settled in the 1870s. Then called Granville, it grew up around the area's flourishing timber industry. In 1886 the town was linked to the rest of the country by the first trans-Canada railroad, and its future as an important commercial population center was established. It was then renamed Vancouver in commemoration of the voyage of George Vancouver, a British sea captain famed for his precise mapping of the west coast of North America.

Between 1792 and 1794 the captain had surveyed the coast from as far south as San Francisco to the southern edge of Alaska. He worked to establish the British claim to some areas on Vancouver Island and confirmed that there was no Northwest Passage, the long-sought water route believed to connect the east and west coasts of North America.

Situated on hills overlooking the harbor with a backdrop of coastal mountain ranges and a mild winter climate, Vancouver is an attractive city for tourists and residents. Its multi-ethnic population, university, and cultural facilities create a sophisticated intellectual environment competitive with Toronto and Montreal.

MELDING OF OLD & NEW

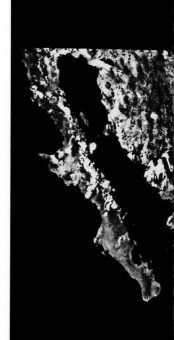

CULTURE AND PHYSICAL GEOGRAPHY HAVE SHAPED THE DEFINITION OF Middle America. A land of volcanoes, deserts, and tropical jungles, it is also a land of lost civilizations and vanished cities, an old land that came to be called the New World. To geographers, Middle America includes Mexico, Central America, the islands of the Caribbean Basin, and the Bahamas. Part of the basis for this definition is Middle America's pre-Columbian and colonial history.

Long before the arrival of Columbus and before Europe's subsequent "discovery" of the New World, several advanced civilizations rose and fell in central and southern Mexico and northern Central America. The Maya, perhaps the most creative people in early Middle America, were unsurpassed in the New World in astronomy, writing, and mathematics. Their culture spread over the Yucatan Peninsula into what would become southern Mexico, Guatemala, and Honduras.

Ruins of Mayan religious or urban centers appear throughout the heartland of Middle America. Over many of the ruins loom large stone pyramids, ascended by steep stairs and capped with temples. The mighty ruins and the immensity of the pyramids evoke a vision of a magnificent civilization. About 600 years before Columbus sailed from Spain, however, the Maya abandoned their centers for reasons unknown, and the culture suddenly declined.

As the jungles encroached upon the abandoned Maya centers, other cultures emerged. The last and mightiest people were the Aztecs, whose empire extended from central Mexico southward into Central America. The capital of the empire was Tenochtitlán, a large urban center on Lake Texcoco that later became the site of Mexico City. About 200,000 people lived in Tenochtitlán—"the place where men become gods"—when the Spanish arrived in 1519.

Shortly thereafter, some 200 years after the building of Tenochtitlán, the city was destroyed. The invaders from the Old World conquered the Aztecs and claimed their land for Spain.

Columbus landed somewhere in the Bahamas when he first arrived in the New World in 1492. Most studies accept San Salvador, or Watling Island, as the most likely landing site. But others argue for no fewer than eight other sites in the Bahamas as the place where Columbus first came ashore. The debate will probably continue until the discovery of solid archaeological evidence—or perhaps Columbus' original log. In the wake of Columbus, the Spanish quickly occupied Cuba, Hispaniola, and Puerto Rico. These islands provided bases for the conquest of Mexico, Central America, and much of South America. The legacy of Spanish colonial rule and its rapid spread from the islands to the mainland creates cultural boundaries for Middle America.

AVHRR MOSAIC AND INSET, DATA COURTESY OF ENVIRONMENTAL RESEARCH INSTITUTE OF MICHIGAN AND THE NATIONAL GEOGRAPHIC SOCIETY. ADDITIONAL PROCESSING BY THE CENTER FOR EARTH AND PLANETARY STUDIES.

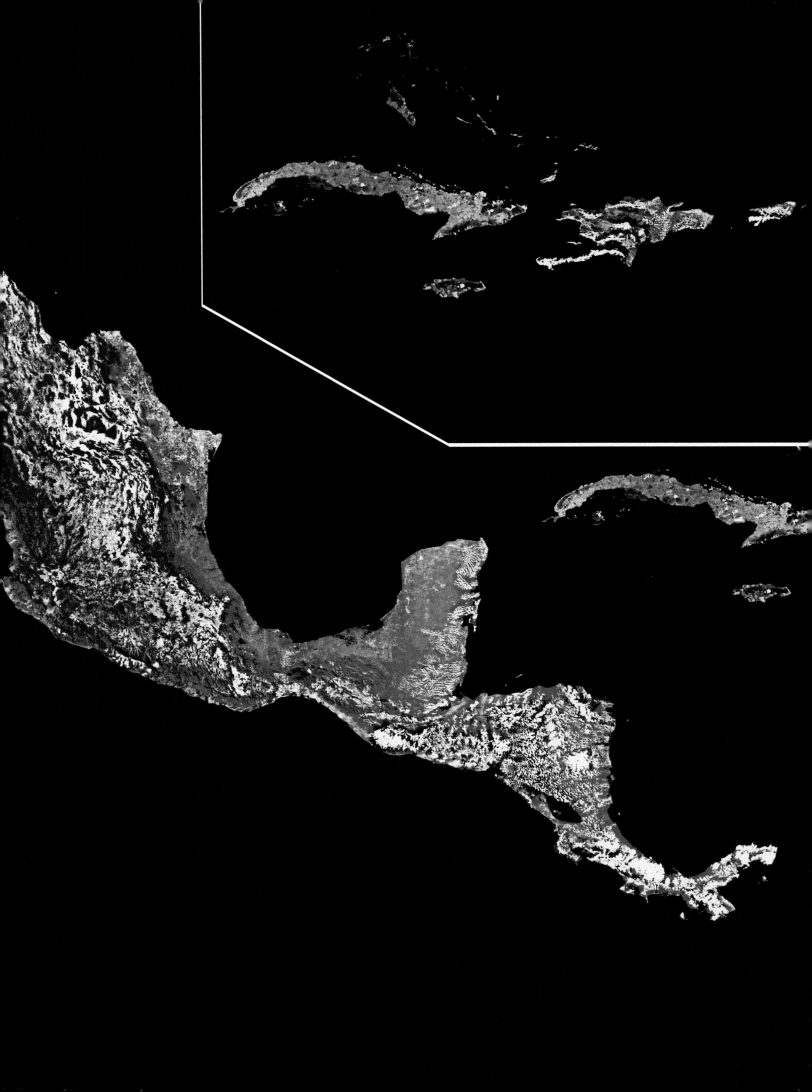

middle america

Northern Europeans also colonized parts of Middle America, most notably the British in Jamaica and Belize, and the French in Haiti. The British and French, along with the Dutch and the Danes, occupied the Lesser Antilles along the eastern margin of the Caribbean Basin. The Spaniards had bypassed these smaller islands as they rushed to the mainland in search of gold.

But the islands' tropical climate and rich volcanic soil gave their European conquerors the opportunity to cultivate a sweeter gold—sugar cane. However, the European appetite for sugar produced one of the most tragic periods in the New World's history. The original inhabitants of the Caribbean islands perished through disease or violence soon after the Europeans arrived, yet sugar plantations demanded a great deal of labor. With no local population to exploit, the European plantation owners began importing slaves from West Africa.

Slavery in the Americas lasted from the beginning of the Spanish conquest until 1886, when it was finally abolished in Cuba. Following the abolition of slavery in the British Empire in 1834, thousands of South Asians immigrated to the islands of Trinidad and Tobago in order to work on the plantations. Over the centuries, the conquered, the conquerors, the slaves, the colonists, and the immigrants have made Middle America probably the most culturally diverse region of the Americas.

Most of the region defined as Middle America lies upon two crustal plates, the North American and the Caribbean. One exception is Baja California, a 1100-km (680-mi.) peninsula that parallels Mexico's west coast and lies on the Pacific plate. The bulk of Mexico lies on the North American plate. Mexico's major landform is the Mexican Plateau, which slopes downward to the north, and is bordered on the east and west by mountains. To the east, the Sierra Madre Oriental range separates the plateau from the Mexican Gulf coastal plain. The Sierra Madre Occidental, a cordillera composed of volcanic material, forms an escarpment above the Pacific coastal plain.

The Isthmus of Tehuantepec, the narrowest part of Mexico, separates the Gulf of Mexico from the Pacific. The Yucatan Channel, which lies between western Cuba and the Yucatan Peninsula, connects the Gulf and the Caribbean Sea. A volcanic line runs down Central America from Guatemala to Panama and indicates the western margin of the Caribbean plate. Its eastern margin is marked by the arc of volcanic islands that form many of the Lesser Antilles.

The tropical climate and the surrounding warm waters of the Caribbean and the Pacific create clouds over parts of Central America during most of the year. Weather in this region is strongly influenced by the northeast trade winds, which carry moisture off the sea and dump rain on windward slopes. During the summer, a band of converging air and low pressure moves north from the equator. This warm, unstable air is a breeding ground for hurricanes that often ravage the region with high winds, storm surges, and flooding.

Much of Middle America is also vulnerable to earthquakes and volcanic eruption. But out of necessity, many Middle Americans live dangerously close to active volcanoes, because some of the region's most fertile land is nearby. The types and patterns of land use in Middle America, as in the rest of the world, are largely shaped by climate, terrain, and soil.

GULF OF CALIFORNIA

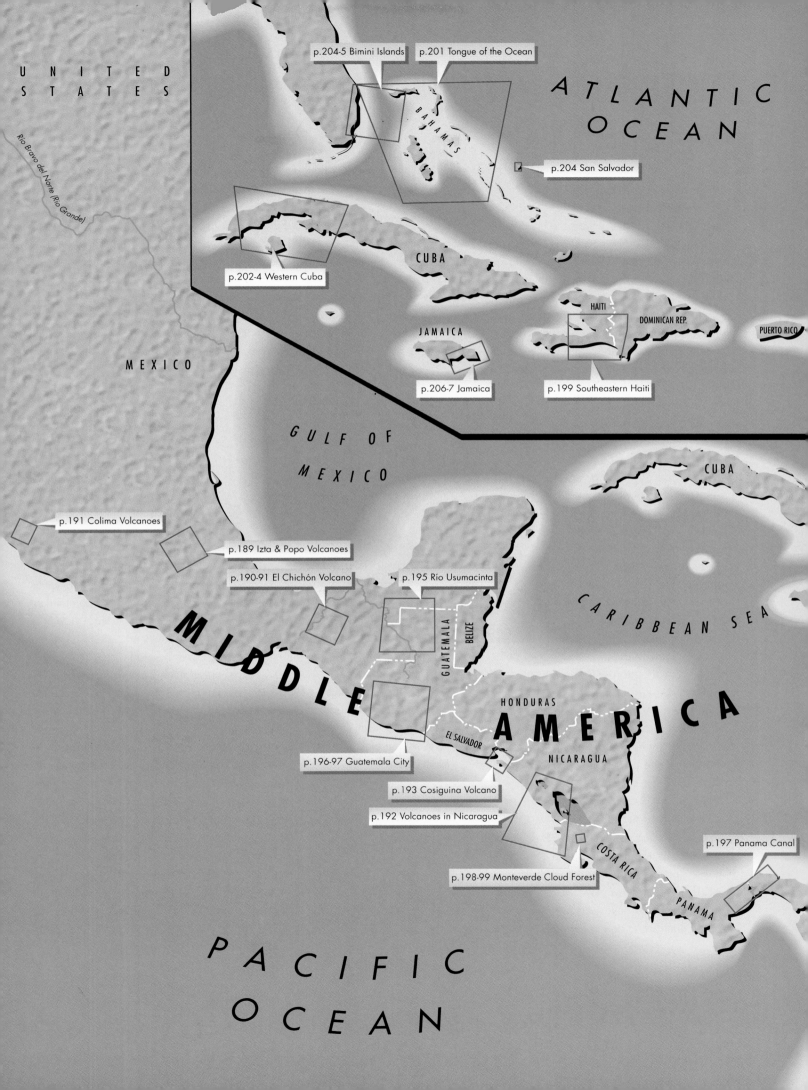

UNITED STATES

Río Bravo del Norte (Río Grande)

MEXICO

ATLANTIC OCEAN

p.204-5 Bimini Islands

p.201 Tongue of the Ocean

BAHAMAS

p.204 San Salvador

p.202-4 Western Cuba

CUBA

JAMAICA

HAITI

DOMINICAN REP.

PUERTO RICO

p.206-7 Jamaica

p.199 Southeastern Haiti

GULF OF MEXICO

CUBA

p.191 Colima Volcanoes

p.189 Izta & Popo Volcanoes

p.190-91 El Chichón Volcano

p.195 Río Usumacinta

CARIBBEAN SEA

MIDDLE

GUATEMALA

BELIZE

HONDURAS

AMERICA

EL SALVADOR

NICARAGUA

p.196-97 Guatemala City

p.193 Cosiguina Volcano

p.192 Volcanoes in Nicaragua

p.197 Panama Canal

COSTA RICA

p.198-99 Monteverde Cloud Forest

PANAMA

PACIFIC OCEAN

Volcanoes of Middle America

FIVE OF THE GREAT, SHIFTING PLATES THAT FORM THE EARTH'S crust converge in Middle America, making it one of the most geologically turbulent areas on the planet. Earthquakes frequently rip through the region, and volcanic eruptions are often extremely explosive. In 1835, for example, a huge eruption, occurring at Cosiguina on the northwest edge of Nicaragua, threw ash more than 1000 km (620 mi.) away (see page 193).

The abundant volcanic activity has produced fertile soils that attract people in search of good farmland. These soils, as well as strategic locations often found near the coast, have drawn people to settle around volcanoes and ignore the danger, sometimes with tragic consequences. One such place is the port of St. Pierre on the island of Martinique. In 1902 one of the most devastating eruptions ever recorded ravaged the island when Mount Pelée exploded, spewing a hot glowing cloud of ash and gas (nuée ardente) that swiftly flowed down onto St. Pierre, killing nearly 30,000 people. Legend says only one person survived—a lone prisoner protected, ironically, in an underground jail cell.

Mount Pelée exploded only a few hours after the volcano of La Soufrière on the island of St. Vincent, 165 km (102 mi.) away, erupted. That eruption killed more than 1,500 islanders. Pelée last erupted between 1929 and 1932 in a series of explosions and hot avalanches that poured down its slopes. People have returned to St. Pierre, living amid ruins that are silent reminders of the power hidden in the Earth's active volcanic zones.

Three major volcanic regions run through Middle America. Crossing roughly east-west through south-central Mexico is a volcanic belt containing such peaks as Popocatépetl and Iztaccíhuatl (opposite page). Around the tall volcanic mountains are thousands of lava flows and cinder cones, volcanic ash deposits, hot springs, and a variety of craters and domes. Lured by fertile volcanic soil and a comfortable climate, people have

The sacred volcano Popocatépetl rises in the sunlight behind the Church of Our Lady of the Remedies, which sits on top of the great, sacred pyramid of Cholula.

settled throughout these volcanic highlands. High basins within the volcanic region include the Valley of Mexico, site of Mexico City. Iztaccíhuatl and Popocatépetl should be visible daily from Mexico City. But the high basin often traps polluted air from innumerable factories and motor vehicles, obstructing the view. Iztaccíhuatl and Popocatépetl have been relatively quiet for some periods and violently explosive at other times. Major eruptions in the past have destroyed portions of each volcano.

Also in the volcanic belt rises one of the Earth's newest volcanoes, Paricutin, which first appeared in 1943. The volcano, which erupted in the middle of a corn field, grew to a height of about 410 m (1,340 ft.) in nine years. Lava engulfed hundreds of homes and destroyed two villages.

Paralleling the Pacific coast is a second chain of volcanic mountains that forms a backbone for the Central American landmass. These young mountains mantle the remains of older volcanic material. The axis stretches from the Mexico-Guatemala border southeast through El Salvador and Nicaragua into Costa Rica. It then continues as isolated peaks into western Panama. Guatemala and Nicaragua (see pages 192-93 and 196-97) have been the sites of many of the most recent eruptions in this volcanic chain. The lack of volcanic activity in Panama made the area a favored site for a trans-isthmus canal. (See page 197.)

The eruptions and other activity near the west coast relate to the subduction of a relatively small plate, called the Cocos, which lies beneath the western edge of the Caribbean plate. On the other side of the Caribbean plate, where it meets the North American plate at the eastern edge of the Caribbean Sea, islands of the Lesser Antilles form yet another volcanic chain. Martinique, the site of the great 1902 eruption, is in the Windward Islands, which form the southern chain of the Lesser Antilles. Volcanoes throughout the region hold the potential for renewed activity.

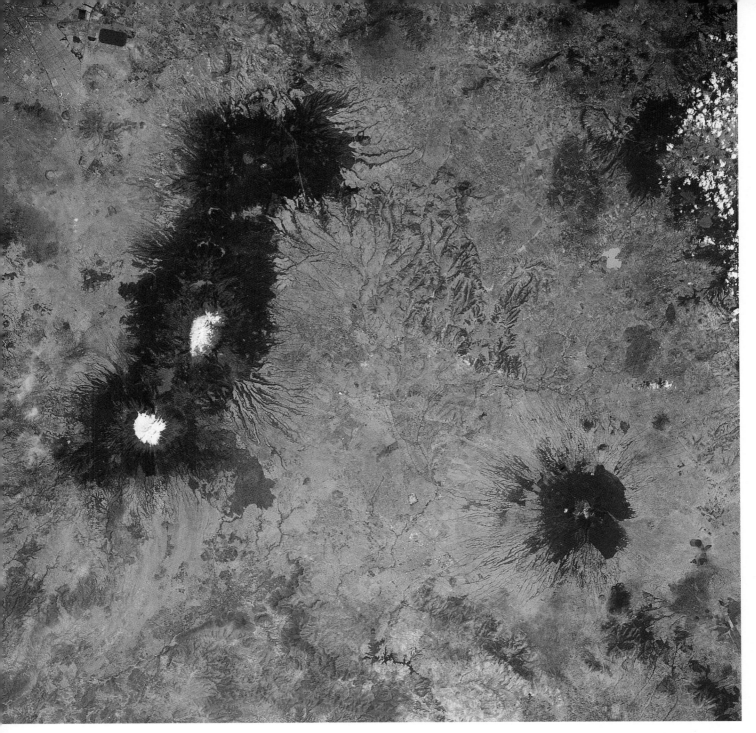

Izta and Popo Volcanoes

Snow-capped Iztaccíhuatl and Popocatépetl tower above the high plains and basins of central Mexico in this image of two of the nation's most renowned volcanoes. Iztaccíhuatl (Izta), the upper volcano in the image, is known as the "Sleeping Lady" or "Woman in White" because the summit of the mountain, fashioned by volcanic eruptions and glacial ice, has the profile of a reclining maiden. Izta, which lies 60 km (40 mi.) southeast of Mexico City, is 5286 m (17,343 ft.) high.

Popocatépetl (Popo), Izta's neighbor to the south, is the second-highest volcano in North America at 5452 m (17,887 ft.). A crater is visible on its summit, and lava flows can be seen on its flanks. Popocatépetl is Aztec for "smoking mountain," a name derived from its occasional volcanic eruptions as well as from the smoke-like plumes that rise from the windswept summit. The volcano is immortalized in Aztec legend as the grieving warrior forever kneeling at the feet of Izta, his deceased sweetheart. Izta has not erupted in historic times, but Popo last erupted in the 1920s.

Another large volcano, La Malinche, rises east of the twin volcanoes. From its summit radiates a well-defined drainage pattern. Forests, easily identified by their dark gray-green color, cover the higher elevations of the volcanic peaks. (D.M.H.)

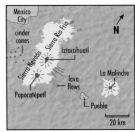

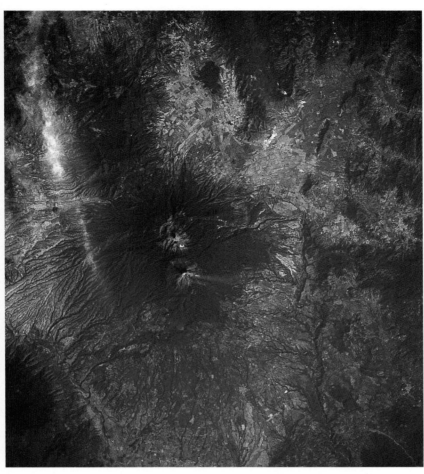

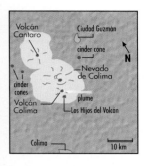

Colima Volcanoes

A chain of volcanoes near Colima, Mexico, stretches across the photo (left), from Volcán Cantaro in the north, through Nevado de Colima and Volcán Colima, ending with a pair of domes, named Los Hijos del Volcán, on the southern flank of Volcán Colima. Los Hijos, "the children" [of the volcano], may mark the inception of the next major volcano along the chain.

Nevado de Colima collapsed three times during its development, forming large calderas that are still visible. The youngest cone, which now marks the highest peak in western Mexico (4320 m or 14,170 ft.), formed within the caldera complex. Volcán Colima, which grew on the southern flank of Nevado de Colima, is one of the most active volcanoes in Middle America. It produced major eruptions in 1818, 1869, 1913, 1961, 1975, 1982, and 1991. A plume can be seen extending from the active crater.

Two major cities are visible, Ciudad Guzmán to the east of Volcán Cantaro, and Colima, at the southern edge. Green fields are also visible. (J.F.L.)

El Chichón

Film sensitive to infrared light that is reflected from vegetation shows the effects of the eruptions of El Chichón (opposite page). For many years, the volcano was dormant and vegetation covered its slopes. Then, between March 28 and April 4 in 1982, came a series of explosive eruptions that threw tons of material into the atmosphere, much as the eruption of Mount St. Helens (pages 168-69) did two years earlier. The violence of the El Chichón eruptions caused small particles to be injected into the stratosphere, where winds distributed them over large regions. Unlike the materials strewn by the St. Helens eruption, the stratospheric particles from El Chichón crossed the equator and continued as far south as São Paulo, Brazil. The new crater (below left) is 200 m lower than the previous one.

In this scene, densely vegetated areas are visible as pink, while more sparsely vegetated semiarid terrains are blue. Ash from El Chichón greatly reduced the vegetation on the flanks of the volcano and has choked the rivers flowing through the area. (J.R.Z.)

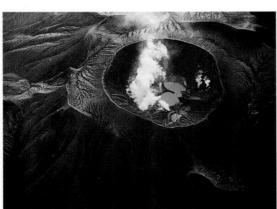

Volcanoes in Nicaragua

This photograph (above) shows an advantage to having astronauts in space: they can recognize and document unanticipated events. Astronauts on the January 1986 Shuttle flight 61-C noticed a long volcanic plume coming from Masaya volcano near Managua, Nicaragua, and were able to record it in a series of photographs. The images provide the only documentation of one stage of Masaya's volcanic activity. They also show a second, smaller volcanic plume from the island volcano of Concepción in Lake Nicaragua.

Lake Managua, with the capital city on its far shore, is bounded by geologic faults along which a devastating earthquake occurred in 1972. The peninsula jutting into Lake Managua was formed by lava and ash from eruptions of the Apoyeque volcano, whose caldera now contains a light-blue lake. Also visible further south on the shore of Lake Nicaragua is a dark mass of lava flows from the Mombacho volcano. (C.A.W.)

Cosiguina Volcano

The long line of volcanoes along the Pacific shore of Nicaragua ends at Cosiguina, a dramatic caldera. In one of the great volcanic explosions of the 19th Century, Cosiguina erupted in 1835. Ash fell as far away as Mexico, Jamaica, and Colombia, and a caldera 2.2 km (1.4 mi.) wide and 500 m (1,600 ft.) deep was formed.

Like some larger volcanic eruptions, the Cosiguina event was followed by cold weather. Volcanoes affect weather by injecting sulfuric gases into the stratosphere. Here the gases are changed into aerosol droplets that are heated by solar radiation, thus depriving the Earth's surface of some of its normal heating. The 1835 eruption has been thought to have led to very cold winters in the U.S., Europe, and Japan, but new information suggests the cold weather may have set in before the volcanic activity.

Also shown in this photograph are relatively new agricultural fields on the flood plain of a small river emptying into the Gulf of Fonseca. Along the northern side of the gulf, concentric lines run parallel to the shoreline. These "strand lines" mark old shorelines. They were formed when sea level was higher or when the land uplifted, raising the old shores above the level of the sea. (C.A.W.)

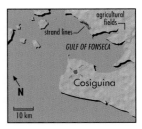

From Aztecs to Agri-business

BEFORE THE ARRIVAL OF THE SPANISH IN THE EARLY 16TH Century, Middle American nomadic hunters and gatherers roamed deserts and grasslands of the north. The Aztecs, ruling from the great city of Tenochtitlán, farmed nearby fertile lake beds. Aztec village councils gave village families the right to parcels of land for as long as the families farmed the land. The system was common throughout most of Middle America. Serfs worked on the large tracts held by Aztec nobles.

The Aztecs cut and piled freshwater vegetation, building a platform, or chinampa, in the lake's shallow waters. They placed mud from the lake's bottom on top of the platform, producing a rich soil in which they planted maize, beans, and chilies. Lake water seeping into the chinampa's base kept the fertile soil moist.

When the Europeans arrived in Middle America, land use drastically changed. Tenochtitlán, seat of Aztec power, was destroyed by the Spanish conquistadors in 1521 and rebuilt as Mexico City. The early Spanish colonists were superb soldiers but were inefficient managers of labor and resources. They introduced the system of encomienda, which gave a colonist, or encomendero, an allotment of land and responsibility for protecting and Christianizing the Indians who lived on it. The encomendero did not receive title to the land. The Indians, who were little more than slaves, worked the fields and mines and paid tribute to the encomendero. Reflecting the outrage of a few Spanish colonists, the Dominican friar Bartolomé de las Casas exposed the abuses of the system. Spain modified the encomienda in the mid–16th Century. But by then the tradition of colonial exploitation had been established in Middle America.

In the 17th Century the British, French, and other Europeans installed a plantation system, based on sugar cultivation and African slave labor, in the Caribbean islands. Plantations were only somewhat more efficient than the hacienda system, which had evolved from the encomienda.

Flourishing from 250 AD to 900 AD, Tikal (above), in the Petén jungle of northern Guatemala, was one of the capitals of the ancient Mayan civilization. Inhabited by as many as 400,000 people, the city spanned 130 sq. km (50 sq. mi.) and had thousands of structures: palaces, marketplaces, temple-pyramids, and causeways. The Maya, who settled Tikal around 600 BC, abandoned it by 900 AD, leaving the city to wildlife and the jungle.

Peasants, or peons, received credit at a hacienda store and eventually ended up in debt to the owner. This practice of debt-peonage gave the hacienda owner, usually a member of a small, elite class of landowners, a steady supply of cheap labor.

The hacienda system lasted into the 20th Century, when revolutions brought some land reform to the region, most notably in Mexico. But the hacienda endures in some form in many Latin American nations.

Like the plantation, today's large commercial farm grows for the export market, specializing in just a few commodities, such as bananas, sugar, cotton, or coffee, and using large amounts of pesticides and fertilizers. Agri-business has left the production of food, particularly grains, to smaller farms and slash-and-burn subsistence farmers. The farmers only clear one small area of forest at a time, just enough to plant crops to feed a family.

The shallow tropical soil is depleted after a few plantings. The farmer then clears a new area. The work of thousands of these subsistence farmers, who are merely trying to feed their families, has been a major cause of deforestation in Middle America. Another has been the so-called Central American "beef boom" of the 1960s and 1970s, largely caused by an increasing demand in the U.S. market.

To meet this demand, tropical forests in Central America were cleared for cattle pasture. From 1981 to 1985, Costa Rica annually lost about 4 percent of its tropical forest, most of it for pasture. Ironically, of Costa Rica's major exports, beef earns the least per unit of production area. In 1984 Costa Rican beef earned about $41 per sq. km (0.4 sq. mi.) used; bananas brought in more than $6,000 per sq. km. The beef boom ended in the 1980s, and it is forecast that demand will continue to decline. Better long-term economic and environmental planning, by all involved parties, is needed if Middle America is going to break the tradition of labor and resource exploitation.

Rio Usumacinta

The middle reaches of the Rio Usumacinta form the border between the Mexican state of Chiapas and Petén, the northernmost department of Guatemala. The contrast between land use in Petén and the Mexican states, particularly Tabasco, is clearly seen in this image. Petén is thinly populated in comparison to the three adjacent Mexican states. In 1985 the population density of Petén was about 3.3 persons per sq. km (1.3 persons per sq. mi.), contrasted with estimated population densities per sq. km for the three adjoining Mexican states: Campeche, 11.7; Tabasco, 51.4; and Chiapas, 33.9.

Higher population densities in these states result in more intensive land use on the Mexican side of the border.

In Chiapas and Tabasco, for example, thousands of hectares of humid forests have been cleared for cattle pastures. These tropical pastures can support more cattle and sheep per unit of area than pre-existing natural grassland ranges farther north. But the ranges on cleared forest lands are overgrazed, eroded, and much less productive after several years of grazing. The trees

of tropical forests in Middle America are also falling to land-clearing homesteaders, loggers, and slash-and-burn farmers who practice shifting cultivation.

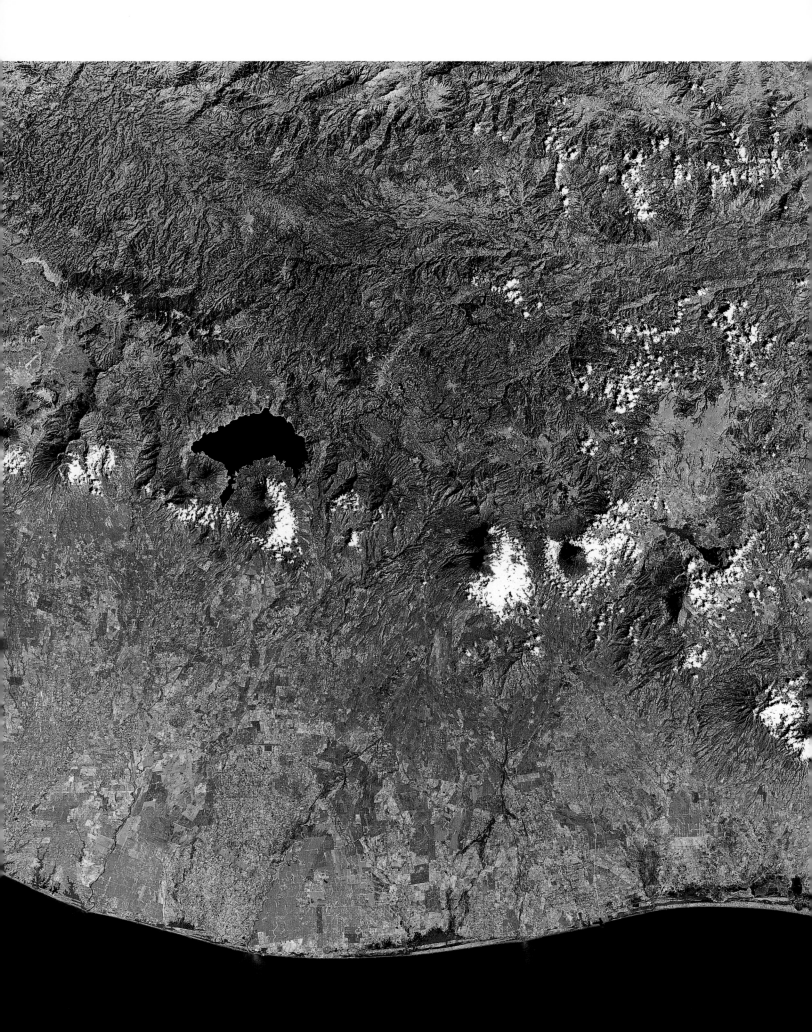

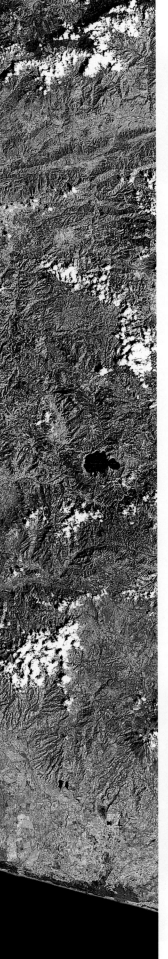

Guatemala City

Clearly visible geometric patterns along Guatemala's narrow Pacific coastal plain show widespread use of this lowland for growing sugar, cotton, and bananas. Coffee, Guatemala's chief crop, is grown on plantations on the slopes of the highlands. The blue color in the valley surrounding Guatemala City (right, center) indicates that this basin has been largely deforested for agricultural and urban development.

Deforestation is becoming a serious national environmental concern. The loss of the soil's protective vegetation increases erosion and run-off, silting up rivers and making them more susceptible to flooding. The clogged rivers impede navigation and hinder irrigation projects. Some of the river-borne sediments are visible as lighter patches where they are discharged into the Pacific.

Guatemala City is in a valley amidst the peaks of the Sierra Madre in Guatemala's central highlands. A line of volcanoes gives evidence of a violent geologic history. Running east to west, north of Guatemala City, the long Motagua Fault is clearly delineated. This area has often been the site of instability and earthquakes. Guatemala City was founded, in fact, after the nearby town of Antigua was destroyed by earthquake in 1773. The city has been repeatedly damaged by quakes ever since. A disastrous earthquake in 1976 killed more than 20,000 people, 1,200 of them in the city.

Panama Canal

Two features of the Panama Canal, Gatun Lake and the Gaillard Cut, show clearly in this radar image (below). The 14.5-km (9-mi.) Gaillard Cut passes through the hills of western Panama. The hills were the greatest obstacle to the canal, and West Indian laborers and North American engineers made the cut using mammoth steam shovels and a rail system for removing earth. The builders also had to overcome diseases, such as malaria and yellow fever, which decimated workers in an earlier attempt to build a canal.

The Panama Canal, begun in 1904 and completed in 1914 by the United States government, uses a system of locks and an artificial lake to cross Panama. Gatun Lake is 26 m (86 ft.) above sea level and was created by the damming of the Chagres River. Ships reach the lake by passing upward through three locks. After crossing the lake, the ships return to sea level through three more locks. Madden Dam on the upper Chagres River controls the flow into Gatun Lake and generates hydroelectric power.

As the waters of Gatun Lake rose, some hills became islands. One, Barro Colorado, is covered with deciduous tropical forest 100 to 400 years old. The isolation of Barro Colorado's plant and animal communities provided tropical forest researchers with a natural laboratory. In 1923 the island became a nature reserve under the auspices of the Smithsonian Tropical Research Institute, a part of the Smithsonian Institution.

Panama City, the capital of the Republic of Panama, is near the Pacific entrance to the canal. The Naos Island Breakwater (bottom center), extending three miles from the mainland to a small group of islands, was built of earth removed from the Gaillard Cut. The breakwater is designed to prevent currents from depositing sediments in the approach to the canal.

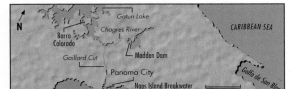

lava flows ✳ —Volcán Arénal

Laguna de Arenal

Monteverde
Cloud Forest
Preserve

N

continental divide

deforested
land

5 km

Monteverde Cloud Forest Preserve

The continental divide runs along the crest of the Cordillera de Tilaran (opposite page, lower left) in northern Costa Rica. The rivers east of the divide drain into the Caribbean Sea; those to the west flow into the Pacific Ocean. Lake Arenal (upper left) lies in the Caribbean watershed, near the 1552-m (5,153-ft.) Volcano Arenal.

To the south of Lake Arenal lies the privately owned, 11,000-hectare (27,100-acre) Monteverde Cloud Forest Preserve, noted for its diversity of wildlife and plant species. This diversity is due to Costa Rica's geographic location between two continents and its mountainous terrain, where plant and animal communities change with elevation. Climate also contributes to diversity by varying with the season and the slope. The Pacific slopes are generally drier than the Caribbean side, where trade winds bring moist air, forming clouds above the forest. Here live more than 100 species of mammals, 120 species of amphibians and reptiles, 400 species of birds, and 2,500 species of plants. To the southwest (lower left) can be seen barren slopes cleared for cattle ranches. Such reduction of habitats will decrease wildlife populations and eventually lessen the diversity of species.

Southeastern Haiti

The effects of deforestation and erosion are clearly visible in this Landsat scene of southeastern Haiti (below). The productive farmlands of the Cul de Sac plain and nearby coastal plains appear bright red, while the slopes of the central plateau and the mountains south of Port-au-Prince, the capital of Haiti, are brown and gray. The border with the Dominican Republic is visible just north of the Étang Saumâtre, where the Haitian side is sparsely vegetated compared to the Dominican side.

Haiti occupies the western third of the island of Hispaniola, with the eastern two-thirds of the island taken up by the Dominican Republic. Haiti consists of a central plateau and two mountainous peninsulas extending to the west toward Cuba. The island of la Gonâve (upper left) lies in the Golfe de la Gonâve, which separates the northern and southern peninsulas. Port-au-Prince (center), the capital of Haiti, is found on the southwest corner of the Cul de Sac plain.

Hispaniola was inhabited by the Arawaks prior to Spanish colonization of the island in the early 1500s. France acquired the western third of Hispaniola in the late 17th Century and developed a colonial economy based on sugar plantations that were worked by slaves imported from West Africa. The colony rebelled against French rule in 1804 and became Haiti. Since independence, agriculture has remained the largest sector of Haiti's economy, employing about 70 percent of the labor force. As Haiti's population increased, the peasant farmers had to cultivate more land. Fragile land on steep mountain slopes was cleared for slash-and-burn agriculture, wood for fuel, and for harvesting timber on a commercial scale.

Forests covered about 75 percent of Haiti prior to colonization. By the late 1970s, less than 7 percent of Haiti remained forested. The soil layer on Haiti's steep mountain slopes readily erodes after the slopes are cleared for planting or from wood gathering. After a few seasons of cultivation, the soil completely washes away, and the land is usually abandoned.

Overpopulation and environmental degradation have combined to make Haiti the poorest nation in the western hemisphere. Because Haiti's land deteriorated and farmers abandoned their fields, the population of Port-au-Prince doubled between 1975 and 1985. Haitians who gave up farming sought jobs in the city, usually in vain.

ISLANDS OF MIDDLE AMERICA

IN THE WATERS EAST OF CENTRAL AMERICA LIE THOUSANDS OF islands, islets, and reefs that make up the West Indies. Stretching from the Bahamas in the north to Trinidad just 16 km (10 mi.) from the coast of Venezuela, the islands mark the boundary between the Atlantic Ocean and the Caribbean Sea. Geologically, the West Indies exhibit a variety of forms. They range from the large mountainous islands of the Greater Antilles to smaller volcanic islands and islands built up from coral and limestones. Many are tiny and uninhabited. Nonetheless, the region is home to more than 30 million people.

Cuba, Jamaica, Hispaniola, and Puerto Rico are the major islands of the Greater Antilles and comprise most of the landmass of the West Indies. The high peaks of an ancient submerged mountain range form the rugged cores of these islands.. All four were visited by Columbus, and during the ensuing Spanish settlement, the populations of Arawak and other inhabitants were all but wiped out. Cuba (see pages 202-4), the largest island in the region, is surrounded by numerous smaller islands and reefs that are also part of the Republic of Cuba. Geographically—if not politically—Cuba is very close to its neighbors. Only 150 km (90 mi.) separate Cuba from Florida. When visibility is good, Jamaica and Hispaniola can be seen from the Cuban coast.

Jamaica (see pages 206-7), the third-largest West Indian island, lies south of Cuba. The name derives from xaymaca, which meant "land of springs" to its early inhabitants. Only 160 km (100 mi.) to the east is Hispaniola, island of two countries: Haiti, which occupies the western third of the island, and the Dominican Republic, on the east. Haiti (see page 199) is made up of two long peninsulas on either side of a deep gulf. Its landscape is largely mountainous, and this rugged terrain extends into the Dominican Republic. The highest and lowest points in the West Indies are found in the Dominican Republic: Pico Duarte at 3175 m

Punta Sardina (above), on the northwestern corner of Puerto Rico, faces the Atlantic. To the west is the Mona Passage between Puerto Rico and Hispaniola, leading to the Caribbean.

(10,414 ft.) and Lago Enriquillo, a salt lake that is 44 m (144 ft.) below sea level.

About 130 km (80 mi.) from the east coast of Hispaniola lies Puerto Rico, an island consisting of a range of central highlands, a northern limestone plateau of karst topography, and low-lying coastal plains. A Spanish colony until 1898, when the United States took control after the Spanish-American War, Puerto Rico has been a semiautonomous commonwealth of the United States since 1952.

The chain of the Bahama Islands lies northeast of the Greater Antilles and stretches for nearly 1000 km (600 mi.). About 700 islands make up the archipelago, but only about 30 are inhabited. The Bahamas (see pages 201 and 204) are flat-lying islands resting in shallow seas. Their name comes from the Spanish term for shallow water. The terrain of the coral islands rarely reaches higher than 60 m (about 200 ft.) above sea level and sometimes rises to only 6 m (20 ft.).

Forming an arc at the eastern edge of the Caribbean are the Leeward and Windward Islands of the Lesser Antilles. The Leeward Islands on the north end of the arc include the Virgin Islands (formed from the peaks of an undersea mountain range) and a chain of coral and volcanic islands. The Windwards to the south are a volcanic chain stretching from Martinique, location of the cataclysmic eruption of Mount Pelée, to the small island of Grenada, site of a 1983 U.S. military invasion. The heat from the volcanic structures fuels hot springs on these islands, and craters and gas vents are common.

A few other small islands and island groups round out the territory of the West Indies. Just to the east of the Windward Islands is Barbados, a tiny, densely populated coral island. The Turks and Caicos islands are flat lands of salt marshes and cactus located at the southeast end of the Bahama chain. The low-lying Cayman Islands, south of Cuba, are surrounded by coral reefs.

Tongue of the Ocean

Numerous small islands mark the highest points of a limestone platform around the Bahama Islands. Most of the platform surface is only a few meters below sea level. But the Tongue of the Ocean, the dark blue water in the center of the image, is a depression within the platform. The sharp contrast in color indicates an area where the ocean is at least one mile

deep. The dark blue of deep ocean water marks the windward (eastern) margin of most major islands of the Bahamas.

The brightest areas indicate oolitic sands, spherical grains of limestone built around shell fragments. Strong tidal currents keep the grains rolling or suspended in the water. Calcium carbonate collects on the growing grain in concentric layers.

Carbonate mud, in places overgrown with interwoven mats of blue-green algae,

covers the top of the limestone platform. Currents in the Straits of Florida carry off the nearby continent's silt and sand, allowing life on the platform to flourish instead of becoming buried beneath a constant supply of mud.

Eleuthera Island appears at the upper right. Eleuthera (from a Greek word for freedom) was named by English religious dissidents who wrecked their ship on sharp reefs here in 1649. For two

years they lived in a cave in the sea bluffs, surviving on plants and animals. The New England Puritans sent supplies to aid the struggling colonists in 1650. In return, the islanders shipped 10 tons of hardwood from the forest on Eleuthera. The light-blue area south of Eleuthera is the Great Bahama Bank. As its loose sand deposits flow with the tides, they form intricate patterns of bright blue to white. (J.R.Z. and R.A.C.)

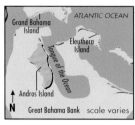

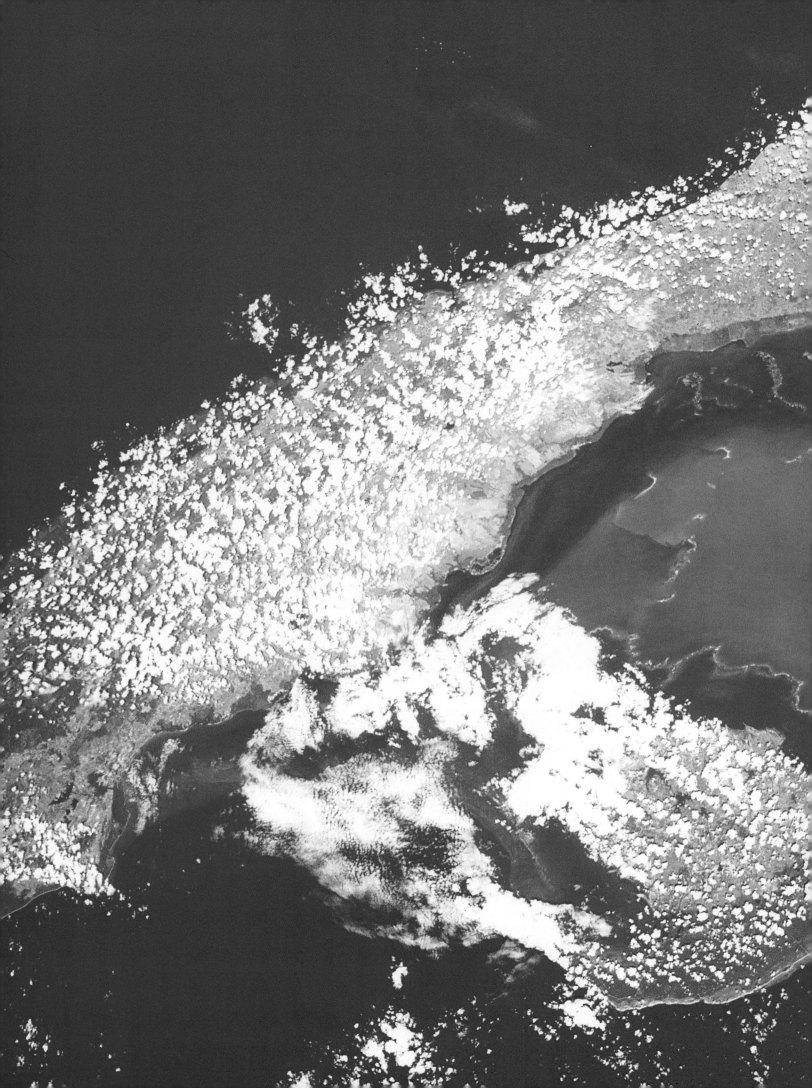

Western Cuba

(previous pages)

Part of western Cuba, centering on the Golfo de Batabanó, appears in this Space Shuttle photograph. On the north coast of the island under the clouds is Havana, the capital. Once a thriving port of call famous for its casinos and nightlife, the city has lost its glitter during three decades of Fidel Castro's authoritarian regime. East of the gulf is the narrow Bay of Pigs, scene of the 1961 U.S.-sponsored failed invasion of Cuba.

To the southwest is the Isla de la Juventud (Isle of Youth). A long stretch of beaches can be seen on its southern coast. The island, formerly known as Isla de Pinos (Isle of Pines), was renamed in honor of the many young people who settled there. Extending eastward from the Isla de la Juventud is a string of narrow islands, the Archipelago de los Canarreos.

Along Cuba's irregular coastlines are numerous bays, harbors, and inlets. The northern coast is rocky, while lowlying marshes and mangrove trees surround the Golfo de Batabanó. Many reefs, cays, and islands dot the gulf's shallow waters.

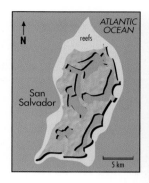

San Salvador

Reefs fringe the small Bahama island of San Salvador (right). Some historians believe that San Salvador is the site of Columbus' first landing in the New World, although the location has been widely debated.

The Bahamian government officially gave San Salvador its name in 1926. Before then, it was often known as Watling Island, after a 17th—Century pirate who had a base there. In the days of Columbus, the Arawaks who inhabited the island called it Guanahani.

A narrow band of sandy beaches can be seen all along the coast. The vegetation across the island (red color) consists mostly of low bushes. The arc-shaped ridges are old vegetated dunes now surrounded by hyper-saline lakes (green).

Bimini Islands

The Great Bahama Bank surrounds the Bimini Islands and extends far beyond their shores. Legend puts Ponce de Leon's Fountain of Youth here, among other places. About 100 km (60 mi.) due west, across

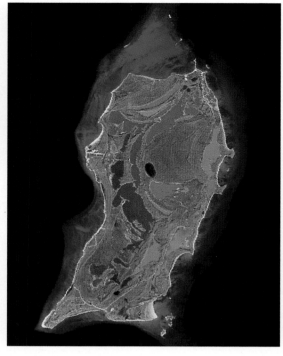

the Straits of Florida, stretches the coast of Florida. A portion of Grand Bahama Island juts into the upper right edge of the image (opposite page).

The Great Bahama Bank is mostly a massive oolite sand deposit. Unstable areas (bright blue) move with the tides and local currents. The dark swath in the straits marks the Gulf Stream. About 99 billion liters (26 billion gallons) of warm, equatorial water move northward past the Biminis each

second. Eventually this water reaches the coast of Scandinavia and helps to keep the climate of Europe warm. Below, a snorkeler finds marine life within arm's reach in the Bahamas. (R.A.C.)

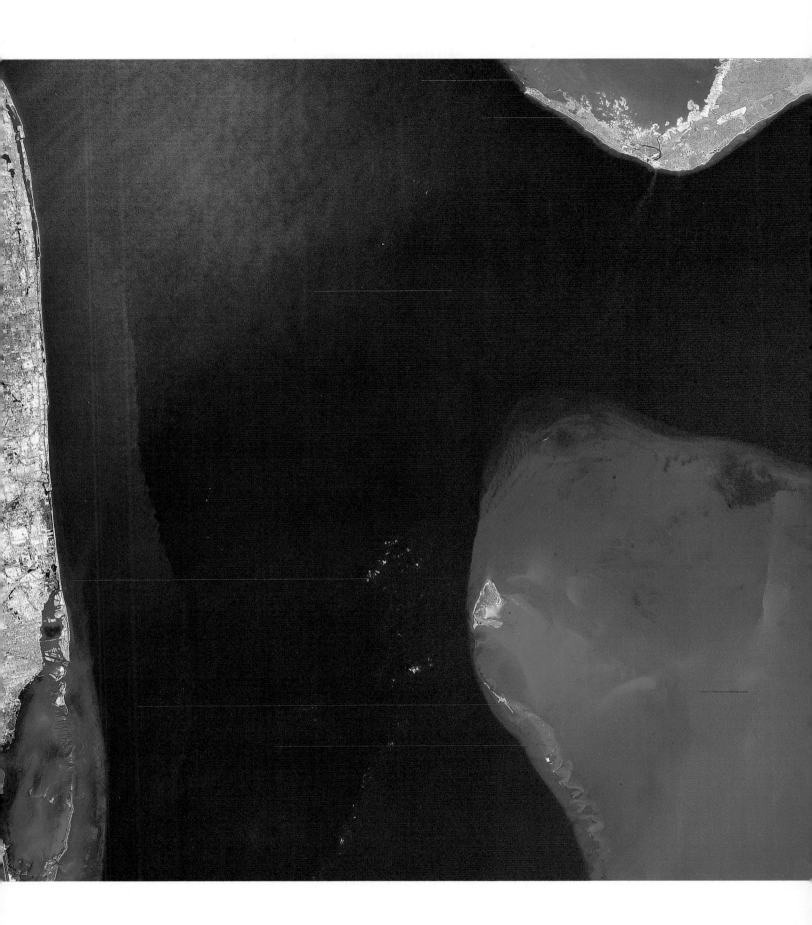

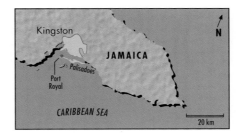

Jamaica

Light-gray areas, indicating urban development, delineate Jamaica's capital city of Kingston (also, photo opposite page) and the narrow 16-km (10-mi.) peninsula, the Palisadoes, which borders Kingston's natural harbor. At the end of the Palisadoes lies Port Royal. A base for pirates, 17th–Century Port Royal was destroyed by an earthquake in 1692. Survivors rebuilt across the harbor. Within a decade the new town, Kingston, was Jamaica's commercial center, and it was designated the capital in 1872. The eastern end of Jamaica is mountainous, like much of the country. These ranges contain Blue Mountain, Jamaica's highest peak (2260 m or 7,400 ft.)

On the narrow flat coastal plain circling the island, Jamaica's main crops of sugar and bananas are grown. In central and western Jamaica the plain gives way to a limestone plateau. The north-central part of this region, "cockpit country," is full of sinkholes and caverns that were formed from the effects of surface and subsurface waters on limestone.

Bauxite, the ore of aluminum, was discovered on the island in the 1940s, and since that time Jamaica has become one of the world's largest producers of bauxite and related products. Most of the bauxite mining operations are in the forested mountain ranges of the interior.

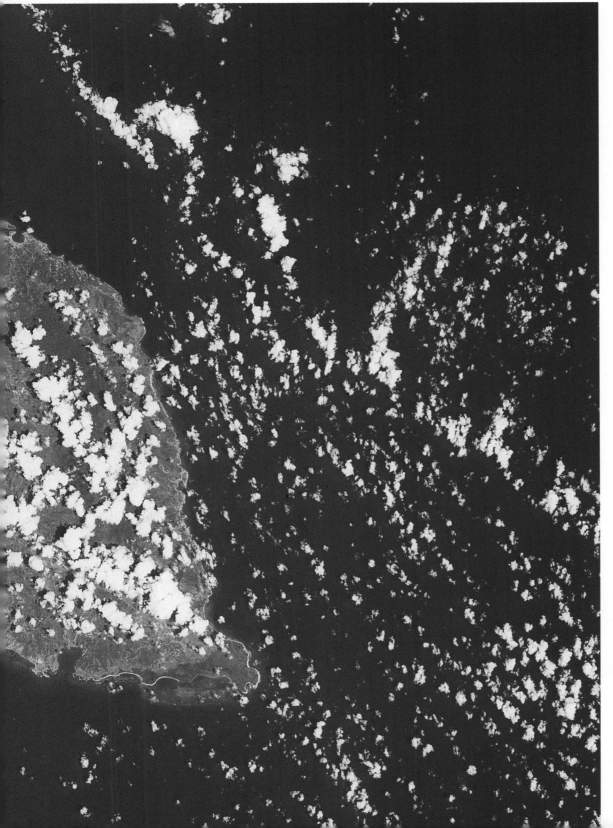

The Hollow Continent

A CONTINENT OF STARK CONTRASTS IN BOTH ITS PHYSICAL AND HUMAN geography, South America has tremendous human and natural resources. Its low population density might suggest potential for growth, but about one half of the continent's lands remain uninhabitable, including the freezing higher elevations of the Andes, the nearly uncultivable tropical forest of the Amazon, and the semiarid Gran Chaco. Contradictions are plentiful: South America's north-central Pacific coast lies in the tropics, but has one of the Earth's driest climates. Great brown rivers meander through a huge tropical forest just a few hundred kilometers across the Andes from this arid coast. In cosmopolitan Rio de Janeiro, shantytowns, or favelas, are scattered throughout the city, not far from expensive apartment buildings, major banks, and corporate offices. The contrasts within South America add up to an inequitable distribution of economic and natural resources.

The most magnificent landform is the cordillera of the Andes, yet this too is subject to instability: below the waters of the eastern Pacific, the Nazca plate is slipping under the western margin of the South American plate. The volcanic eruptions and earthquakes that are common in the Andes stem from this subduction process.

To the east of the central Andes, the Earth's largest tropical forest, or selva, occupies most of the Amazon basin. Evergreen hardwoods 40 m (130 ft.) high form a thick canopy far above the forest floor. Below is not a thick impenetrable jungle, but rather, open space that receives no direct sunlight and is home to almost one half of the Earth's plant and animal species, the planet's largest storehouse of biological diversity.

Biodiversity refers not only to numbers of species, but also to genetic diversity, which enables a species to adapt to a changing environment. Deforestation of the Amazon basin has been recognized, almost universally, as a danger to the Earth's biodiversity, yet economic and political pressure to clear the forest for cultivation or ranching persists and dominates policy decisions regarding land use in Amazonia.

The Amazon basin is enclosed by the Andes to the west, the Brazilian Shield to the south, and the Guiana Shield to the north. The Brazilian and Guianan shields are composed of deeply eroded ancient crystalline rock, and the highlands they form are covered with tropical savanna, known locally as cerrados. Poor soil of the cerrados has limited agriculture, so cattle grazing is the most common activity on the highlands. Much of South America's bulge into the Atlantic Ocean is the eastern Brazilian Shield. Known as Brazil's Nordeste, it is drought-prone and heavily populated. Nordeste's large impoverished population is a legacy of the labor-intensive sugar and cotton cultivation of the previous centuries.

COURTESY OF EROS DATA CENTER,
NATIONAL MAPPING DIVISION,
U.S. GEOLOGICAL SURVEY,
DATA COURTESY OF NOAA

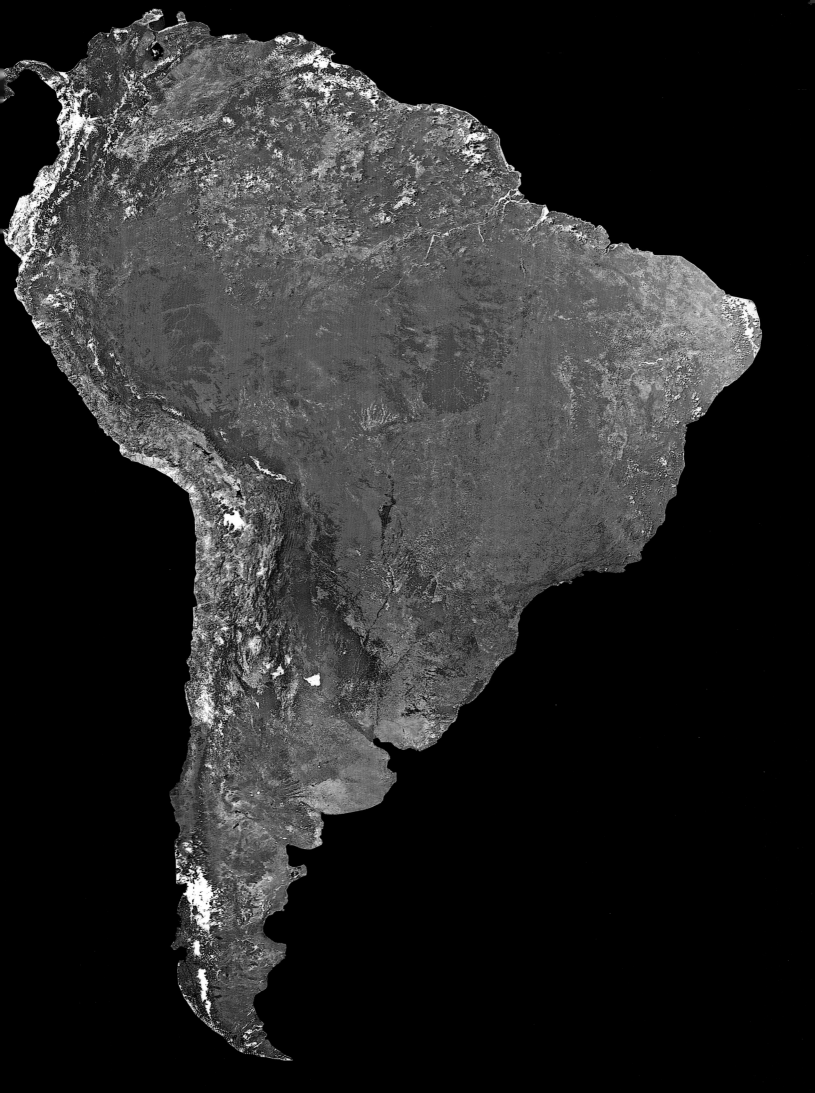

South of the Brazilian Shield lie the Paraguai and Paraná drainage basins. Both systems drain the Brazilian highlands and flow to the South Atlantic through the Río de la Plata estuary. South and west of the estuary the almost-flat Argentine Pampa gradually rises toward the Andes to the west. A rich grassland covering thick, loose sediments eroded from the mountains, the Pampa grades from grassy plain into temperate desert in Patagonia to the south. Tierra del Fuego, the "land of fire," and a number of nearby smaller islands are actually a part of the Andes that curves eastward to form the southern tip of the continent. Thus, the coastal archipelagos of southern Chile are submerged mountains. The southern Andes are lower than the northern cordilleras, and these mountains have glaciers, some of which reach down to sea level. Fiords and a cool, wet marine climate make this region unique compared to the rest of South America.

North of Chile's temperate middle region, the central Pacific coast has one of the driest climates on Earth. The effect of an upwelling of cold water near shore, the northward flow of the cold Peru current further offshore, and the barrier of the Andes combine to create a narrow coastal desert in northern Chile and Peru. Peru's largest region, however, is the selva and it includes the great forests east of the Andes in the Amazon basin. The sierra region of Peru generally lies above 2000 m (6,500 ft.), and ranges from cultivated areas on the Altiplano to the snow-covered peaks of the cordilleras.

Ecuador's coast, by contrast, is not arid and has grazing lands and coffee, cocoa, and banana plantations. The sierra narrows in Ecuador as the Andes continue across Colombia, where mountains become three roughly parallel cordilleras. Like Peru and Ecuador, the eastern part of the country is selva. A grass-covered plain, known as the Llanos, extends into Venezuela from north of the selva and east of the Colombian Andes.

Eastern Venezuela is a highland region that is part of the Guiana Shield. Tepuis, the region's high table mountains, have vertical sides of almost 1000 m (3,300 ft.). On the broad tops are assortments of rare species that evolved in isolation. Draining the Llanos and the western Guiana Shield is the Río Orinoco. To the east of its delta are the three smallest countries in South America, known as the Guianas. Guyana is a former British colony that specialized in sugar production, as did the former Dutch colony of Suriname. French Guiana is not an independent country, but is an overseas department of France, famous for its space center at Kourou and its historic, miserable penal colonies, such as Devil's Island. The hot, wet Guianas have the lowest population density in South America.

South America is sometimes described as the "hollow continent" because it is empty in the middle: three quarters of South America's population is urban, and most of the continent's major cities are coastal. Urbanization is overwhelming the infrastructure and decreasing the environmental quality of South America's cities. In a trend opposite to urbanization, but also environmentally degrading, migrants are clearing tropical forests for cultivation or grazing. It has been said that the root cause of South America's problems is its colonial legacy of resource and human exploitation. Perhaps the South Americans will end this legacy as they enter a new millennium with the lessons of the past 500 years and make the most of their unique blend of American, European, African, and Asian cultures.

PACIFIC OCEAN

VENEZUELA

Río Orinoco

COLOMBIA

GUYANA

SURINAME

FRENCH GUIANA

p.236-37 Mouth of the Amazon

ECUADOR

p.218-20 Manaus

Río Amazon

PERU

SOUTH AMERICA

BRAZIL

p.213 Lima

p.214-15 Cuzco

p.226-27 Lake Titicaca

p.232 Bolivian Lowlands

p.217 Brasília

p.238 Nazca

p.233 San José de Chiquitos

BOLIVIA

p.228-29 Colca Canyon

p.234-36 Lands Wet and Dry

p.220-21 Rio de Janeiro

p.225 Bolivian Altiplano

p.224-25 Atacama Desert

PARAGUAY

Río Paraná

p.231 Cataratas del Iguazú

URUGUAY

p.239 Río Paraná

p.216-17 Buenos Aires

CHILE

ARGENTINA

ATLANTIC

OCEAN

The Overpowering Lure of the Cities

IN LESS THAN A CENTURY SOUTH AMERICA HAS EVOLVED FROM a continent with a dispersed agricultural population to one that is more urbanized than any other continent. In 1990 about 76 percent of South America's population was urban, compared to 73 percent in Europe. (Asia is the least urbanized of the continents, with 30 percent of its population living in towns or cities.)

Following the Spanish conquest of South America, the cities of South America—once centers of political power and religious ceremony—became outposts of a global empire. The conquistador Francisco Pizarro, averse to the lofty, thin-aired Inca capital of Cuzco, established Lima as his capital. The Spanish built a port at Callao, about 12 km (7.4 mi.) from Lima, for shipping gold and silver to Spain.

Cuzco eventually slipped to the status of a provincial capital. Overland trading routes, bypassing Cuzco, connected Lima, the new economic and political center of colonial South America, with the silver-mining boom town of Potosi in the Bolivian mountains. In 1600, Potosi had a population of 120,000 and was the largest city in the Americas. For more than a century silver flowed steadily from the mine at Potosi to Spain, providing a major portion of the nation's wealth.

Government-backed merchant guilds monopolized trade with the Spanish colonies. The monopoly, which lasted until 1778, concentrated merchant and trading activity at a few major South American ports and established the primacy of coastal cities.

The Spanish also maintained tight control over the layout of cities. A grid pattern of streets—with a central plaza for church, city hall, and governor's residence, or palace—was mandated in 1573, and set a pattern repeated on all scales, from large coastal cities to remote Andean towns. In the pre-Columbian city, the population consisted of an urban elite. But in the Spanish colonial city lived craftsmen, soldiers, merchants, and administra-

S hantytowns, or favelas, crowd any unused and accessible land in Rio de Janeiro (above). The favelas are often the final destination for many Brasileiros who, in search of a better way to provide for their families, migrate from rural areas to Rio, a city of 13.5 million people.

tors. Social status could be measured by how close a house was to the central square. Near the city's edge, where the grid pattern broke down, the poorest urban dwellers settled. This pattern persisted until the arrival of industry in the 20th Century.

As recently as 1940 the continent's rural population stood at 70 percent. Then, as overall population grew, migration from rural to urban areas began. Farm workers left the countryside because the hacienda owners still controlled the best lands, squeezing out the growing rural population. Most migrants headed for coastal cities because, as longtime centers of power, they had attracted investments in new business and industry.

Migrants have moved in an oft-repeated pattern: one family member moves to the city, finds a job and a place to live, and then sends for other family members. But the supply of workers for good jobs has far exceeded the need. Therefore, for many migrants, the only jobs available in the informal economy are on the street or in markets.

The old colonial pattern, which put the most distinguished homes in the city center, has changed as the urban elites have moved to exclusive suburbs. Another form of suburb is the shantytown that springs up on unused land at the edge of the city as well as in empty lots or in alleys. The squatters' claim to unused lands has the tacit approval of the government, which recognizes its inability to house the poor.

Urban populations in South America have surged dramatically over the last few decades as a result of population growth and the migration from rural areas. In the 1980s economic problems, by some estimates, set South America back ten years. Financially strapped governments could not improve or expand the urban resources needed to accommodate the steady stream of people into the cities. The result is a deteriorating urban environment that suffers from polluted air and water, overcrowding, crime, and unemployment.

Lima, Peru

The gray sprawl of Lima spreads upward from the coast in this image of Peru's capital. Of the country's 22.3 million people, about 5.6 million—25 percent—live in greater Lima. In population, Lima is twelve times the size of Arequipa, Peru's second-largest city. Lima dominates most of Peru's political, cultural, and economic activities.

Lima's primacy dates back to the founding of the city by the conquistadors in 1535. As the capital of Viceroyalty of New Spain, a territory that originally included all of Spanish South America, Lima, and its port, Callao, had a monopoly on trade between the continent and Spain. Lima's population has grown dramatically since the turn of the century, when the city had 150,000 inhabitants. Today, little housing is available to the poor from the rural highlands. Since about 1950, squatters have settled on empty desert land and slopes near the city's edge. A squatter settlement usually begins with an overnight invasion of a parcel of land during which hundreds of squatters build shelters out of straw mats, cardboard, and scrap lumber. The crude shelters do not need good roofs in Lima's arid climate. For five to ten years, the invasiones lack basic services. But the squatters make improvements, and the city eventually provides public services. At this point, the settlement becomes a legitimate pueblo jovene, or young town.

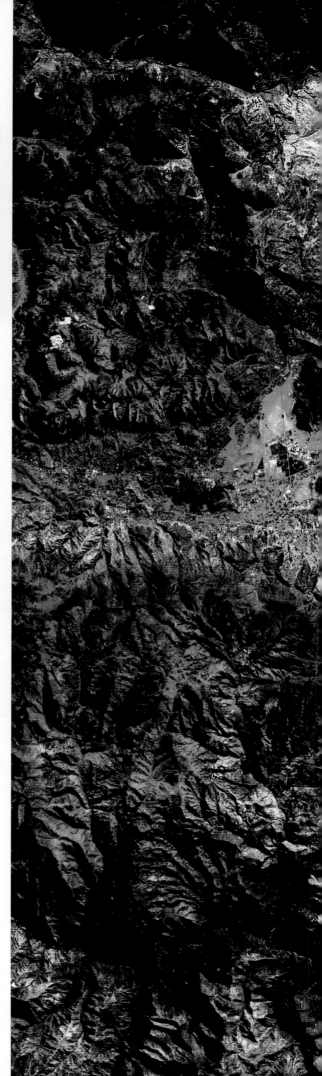

Cuzco, Peru

Survivor of a vanished civilization, Cuzco emerges from the Andean heights of Peru in this SPOT image. It is one of South America's very few pre-conquest cities. After the conquest, the Spanish could not impose their grid street plan on Cuzco because the city's Inca palaces and temples were too massive to be destroyed. The Spanish built on top of the Inca stonework, and many buildings in Cuzco have walls on their ground floors composed of tight-fitting, uncemented stone blocks (above). When the earth quakes in this active seismic region, the Inca stonework survives, while buildings of colonial and more recent eras often collapse.

The great Inca fortress called Sacsayhuaman, or Many-colored Falcon, overlooks the city to the north. Inca dams, aqueducts, and shrines dot the hills surrounding Cuzco.

An Inca myth says the Sun sent his son Manco Capac and Mama Ocllo, the daughter of the moon, his wife, and sister, to a dark, chaotic Earth. They arrived by rising from the waters of Lake Titicaca and, in search of a site for their kingdom, traveled northwest to the Río Huatanay valley. Here Manco probed the earth with his staff, found thick, fertile soil, and named the place Cuzco, "navel of the world." Cuzco became the Incas' center of power, religion, and culture.

The Inca Empire probably began as a confederation of city-states that included Cuzco and the towns in the valleys of the Río Vilcanota and the Río Urubamba, such as Pisac and Ollantaytambo. Local gods were assimilated into the Inca pantheon, and the sons of local chiefs were taken to Cuzco for education and later returned to rule their home provinces. A network of roads connected Cuzco with the rest of the em-

pire. At its zenith, the empire extended from central Ecuador south to central Chile, and from the Pacific coast of South America to the eastern slopes of the Andes.

The Inca people were subject to state control of almost every facet of life, from the type of clothing they could wear to the size of their homes. Compulsory labor built public works and tilled agricultural lands, providing food for all. A reliable source of food was one of the benefits of life under the rigidly enforced caste system and an efficient, stable state.

But in 1532, when civil war split the empire, the conquistador Francisco Pizarro marched unopposed into the Andes from the Pacific coast. His seasoned force of 150 soldiers included veterans of the conquest of the Aztecs. Within a year Pizarro's force occupied Cuzco and installed a puppet Inca ruler, who later led an unsuccessful rebellion against the Spanish.

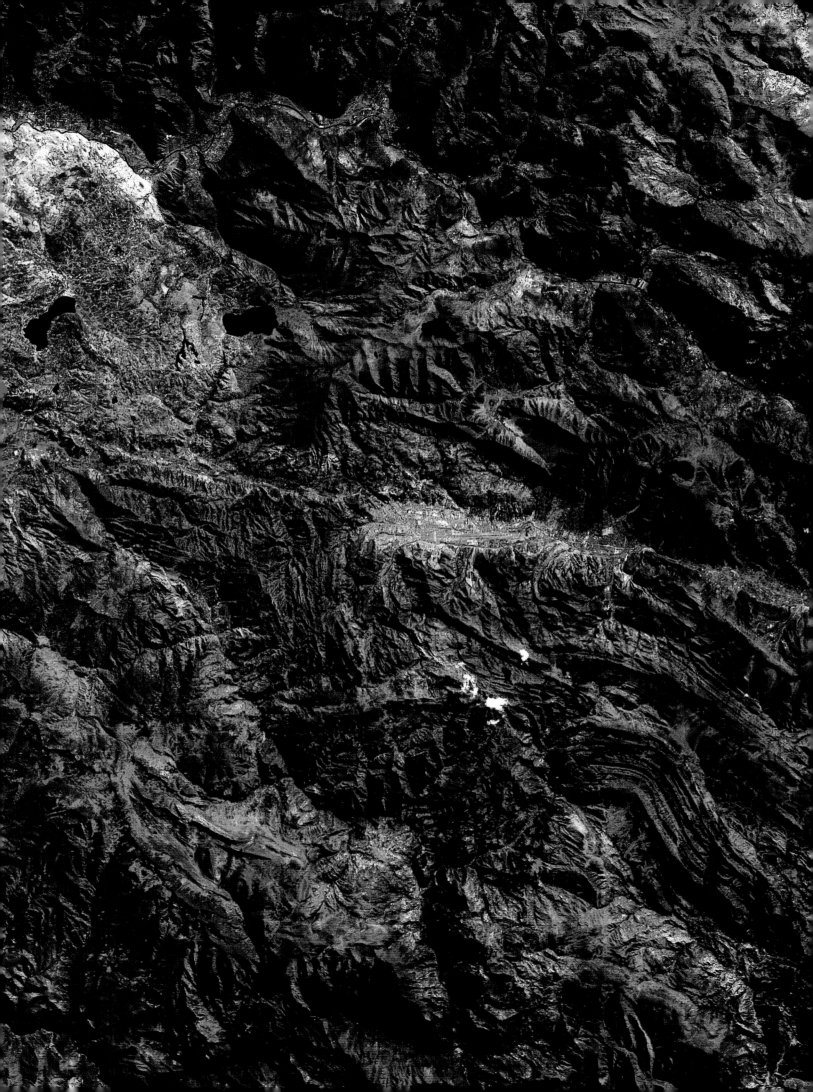

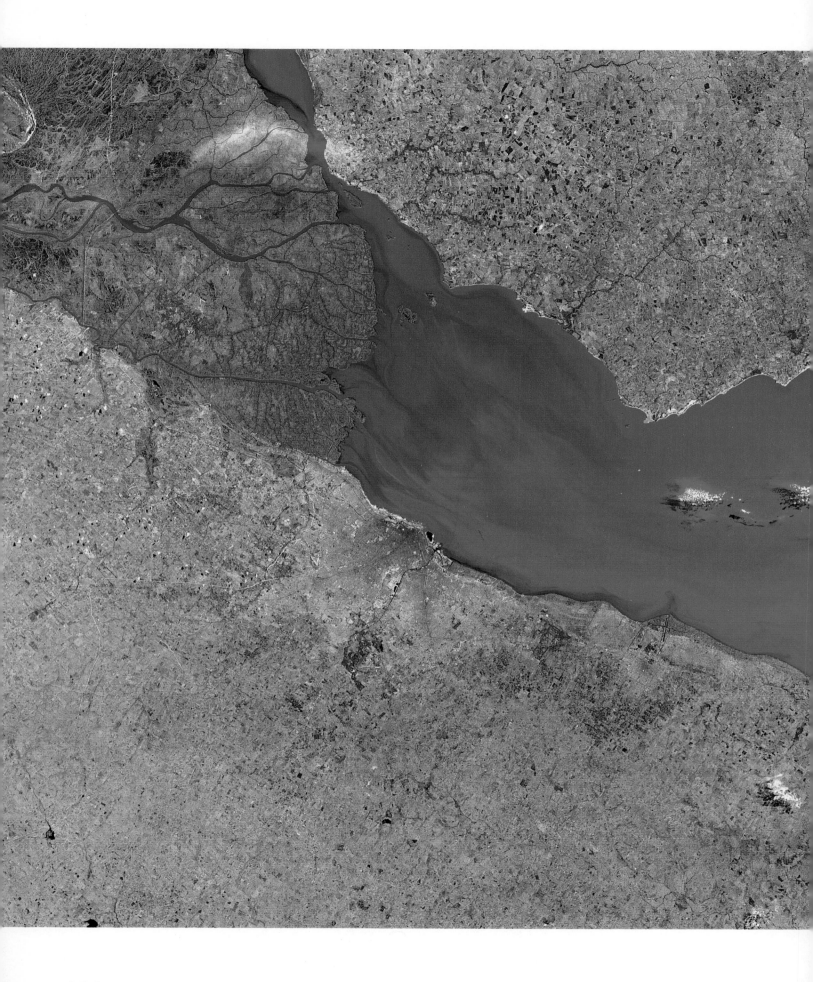

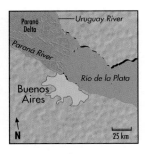

Buenos Aires

A light-blue urban sprawl marks Buenos Aires, a city of about 10 million people that is the capital of Argentina and is the Western Hemisphere's third-largest metropolis. The city is made up of numerous diverse barrios (neighborhoods), many of which date back to the 19th Century. The centro (city center) marks the oldest part of Buenos Aires and contains major government and commercial facilities.

Buenos Aires lies on the south bank of the Río de la Plata (River Plate), an estuary at the confluence of the Paraná and Uruguay rivers. Estuaries form near the mouths of rivers and are zones of transition, where the rivers' fresh water mixes with salt water as the river empties into the sea.

The Río de la Plata is bordered on the northwest by a vast delta formed from sediments transported by the swiftly flowing Paraná River. These sediments (indicated by the mottled blue and green patterns visible in the estuary) clog the waterway, making the route to the port of Buenos Aires shallow and marshy. Because the port is shallow, it has been extensively developed and is dredged continually to maintain ease of access for even the largest commercial vessels.

Buenos Aires is on the northeastern edge of the Pampa, an extensive flat and fertile grassland where the rich soil supports diverse farming and provides pasture land. Spanish explorers first settled here in 1536 but were driven out by native inhabitants. Not until 1580 was a permanent European settlement established. The Europeans brought to the region the livestock that was the foundation for Argentina's cattle industry, and the grasslands that surround the city and reach south—along with a temperate climate and good water supply—were ideal for cattle raising.

Thanks to numerous waterways and fertile lands, Buenos Aires grew into a successful commercial hub by the mid-1800s. It is particularly noted as a center for the exporting of wool, meat, hides, grain, and dairy products. Railroads connect the capital with many regions of the country. Most of Argentina's industry, commerce, government, educational, and cultural activities are conducted in and around Buenos Aires.

Brasília

The intriguing bird-like shape of Brasília appears in this view of Brazil's capital. Carved out of a wilderness, the city was laid out with the residential areas running along the "wings" and the government and cultural centers extending down the central axis. Lake Paranóa, as newly built as the city, partially surrounds Brasília.

Brasília is in the south-central interior of the country, about 960 km (600 mi.) northwest of Rio de Janeiro, the former capital. Brazil, the largest country in South America, covers an area greater than that of the coterminous United States. But most Brazilians live and work in a relatively small area along the coast. To alleviate overcrowding in the coastal area, to open up the vast interior for development, and to establish a centralized government more accessible to the inland states, President Juscelino Kubitschek launched a design program for a new capital in 1956. When the design was approved, construction began on Brasília, which opened in 1960. The city is part of a larger Federal District that includes several satellite cities. Although the government is the major employer in the region, there is also some light industry.

The city is at the source of several rivers that provide broad access across the country. But at the start of construction there were no roads to the area, and much of the materials and equipment had to be flown in. Building Brasília in a mere four years was an impressive feat. Kubitschek hoped that the building of Brasília would spur rapid development around the city and throughout the country. These expensive development programs provided many much-needed jobs but also greatly increased Brazil's inflation and foreign debt.

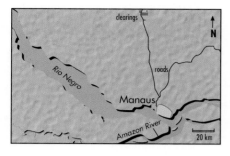

(previous pages)

Manaus, Brazil

Two rivers meet but do not immediately merge in this image showing the joining of the Rio Negro and the Amazon River near Manaus, Brazil. Black waters of the Rio Negro and the muddy, silt-laden waters of the Amazon meet but do not mix, and continue to flow together down the same stream bed, with their contrasting colors evident, for some 80 km (50 mi.) past Manaus.

The capital of Amazonas, Brazil's largest state, Manaus is home to more than a million people, the largest population center in an otherwise sparsely settled and dense rain forest.

Around the turn of the century, a virtual monopoly on rubber trade made Manaus an active center of commerce, industry, and the arts. When the Brazilian rubber trade collapsed, Manaus stagnated.

In the 1960s, after the government made Manaus a duty-free zone, the city revived. Farming and mining soon ate into the forest. In this image from 1977, the beginnings of the deforestation are visible, especially in clearings along roads.

Rio de Janeiro

Numerous vessels dot the Baía de Guanabara, harbor of Rio de Janeiro (right). The thin line spanning the bay is the Presidente Costa e Silva Bridge, connecting Rio to the suburb of Niteroi and Brazil's largest shipyards. Galeão International Airport on the Ilha do Governador is also visible in the Space Shuttle image, as are the famous beaches of Copacabana and Ipanema. At the mouth of the bay on a small peninsula lies the steep rounded cone of Sugar Loaf Mountain (below), a granite pinnacle whose shape was formed by the erosive forces of the tropical environment.

Rio's name dates back to the beginning of the 16th Century, when Portuguese navigators, thinking the bay was a river mouth, called the region Rio de Janeiro (River of January). The city grew up from a town around a citadel built in the 1560s as a base from which to drive French colonists from the area.

Spurred by discoveries of gold and diamonds, 18th–Century Rio's population grew rapidly, and after Brazil won independence from Portugal in 1822, Rio became the national capital. It remained so until 1960, when the seat of government moved to Brasilia.

Today Rio is a city of both skyscrapers and shantytowns, housing a population of more than 5 million, nearly half of whom migrated from rural areas of Brazil. Although Brasília is now the capital, Rio remains the cultural and artistic heart of the country.

The Andes—Giant Range in a Restless Land

THE LONGEST MOUNTAIN RANGE IN THE WORLD, THE ANDES outline the west coast of South America, stretching from Venezuela in the north to Tierra del Fuego at the continent's southernmost tip. Numerous volcanoes and deep faults in the Andes indicate a long history of geologic instability that continues to today.

The Andes are not actually a single range but a series of parallel chains, together called a cordillera, interrupted by flat-floored basins perched at high altitudes. The forces that formed the Andes cordillera are complex and represent geologic upheavals of different magnitudes through the ages. Broadly, the structure of these mountain ranges has to do with their location on a crustal plate boundary along the Pacific rim. Here ocean crust is subducted under the South American plate. Here, too, violent geologic activity sometimes occurs.

In Colombia, the northern Andes have three ranges, one of which extends into Venezuela. The central part of the cordillera, stretching from Ecuador in the north to central Chile and Argentina in the south, contains the highest Andean peaks. In Ecuador, two branches of the Andes surround the Quito basin, site of the capital, Quito.

Numerous volcanic peaks, some of them active, surround this region, which is about 2900 m (9,500 ft.) above sea level. The ancient city of Quito, which dates back to before the 11th Century, lies near the foot of the Pichincha Volcano. This volcano is active to this day, venting gases and steam and giving off small explosions. Pichincha's last major eruption, in 1660, deposited 40 cm (16 in.) of ash on Quito and produced glowing avalanches of ash and gas (núees ardentes). The area has also been repeatedly ripped by earthquakes.

The Andes cordillera, about midway along its course, makes a sharp bend that runs parallel to the South American coast. In this region of Peru, Bolivia, and Argentina is found the high plain called the Altiplano (see page 225). An arid region characterized by volcanic structures and lava flows, the Altiplano rests between two Andean ranges. In this area is one of the world's greatest concentrations of volcanoes.

A prehistoric sea evaporated, leaving behind a large salt plain in southwestern Bolivia. Though dry and treeless, this land is inhabited by South American Indians, mainly the Aymara, who, after being conquered by Pizarro in 1538, continued their agricultural and pastoral culture. Above, llamas, which are used to transport salt blocks to market, rest at a brackish watering hole.

In central Argentina along the border with Chile, is Aconcagua, the highest peak of the Andes and, at 6960 m (22,830 ft.), the tallest mountain in the Western Hemisphere. Ferocious winds rake Aconcagua, a challenge to mountain climbers. Dozens have died on its freezing flanks. South of Aconcagua mountain heights are lower as the Andes develop into a narrow band with scattered volcanoes that are both active and inactive. Glaciers and ice caps appear among the peaks of southern Chile.

The Andes have long been known for their abundance of ore deposits. The gold and silver found there fueled the Spanish conquest and occupation. Incas called gold the "sweat of the sun," and they called silver "tears of the moon." In search of gold and silver, conquistadors plunged into the wild and hostile Andes. Earthquakes destroyed Spanish settlements, and Indians struck down many invaders. Legend says fierce Araucanian Indians killed one treasure hunter by pouring molten gold down his throat—because, they said, it was gold he had been seeking.

In the 20th Century, gold and silver production has diminished. But other ores still drive a major part of many South American economies. In the middle of the 19th Century, copper, tin, and iron ore industries were developed in the Andes. By the late 1800s Chile led the world in copper production. One of the world's largest open-pit copper mines is found today at Chuquicamata, Chile.

In spite of high altitudes and harsh conditions, the Andes gave rise to agricultural civilizations that utilized extensive irrigation networks and were ruled from cities that served as administrative and religious centers. Prior to the Spanish conquest, populations numbering in the millions thrived on specialized crops developed to adapt to the dry climate, high altitudes, and extremes of temperature characteristic of the Andean environment.

Shape of the Land

Satellite imagery combined with topographic data produced this image, which shows much of South America in one view. This stunning portrait of a continent dramatically illustrates the contrast between the rugged terrain of the west coast and the lowlands to the east.

The peaks and high basins of the Andes, rising along the west coast, form a challenging barrier to trade and to cultural communications. For this reason, Andean populations through history often have been isolated from outside influences. So inaccessible are the homelands of some Andes peoples that some aspects of early cultures—such as language, customs, and dress—have been preserved intact or only slightly modified. Highland Indians of Peru, for example, still speak a language that was spoken by Inca ancestors.

Atacama Desert

One of the driest areas on the Earth, the Atacama Desert appears in this view of a segment of the northern Pacific coast of Chile (opposite page). Offshore cloud banks signal a cause for dryness: potential moisture over the Pacific is lost before reaching land due to the upwelling cold waters of the Humboldt Current. Sparse vegetation survives on the coastal cliffs, where daily fog, the camanchacha, provides limited water, which sustains a delicate coastal ecosystem.

Antofagasta, which lies near the northern edge of the image, is Chile's fifth-largest city (population 183,000) and a port for exporting ore from desert mines. The city gets water by aqueduct from the high Andes some 200 km (120 mi.) east.

Dark colors of the coast ranges are largely due to old volcanic rocks. The region's scant moisture evaporates in low-lying depressions, where various salts accumulate. Small salt flats (salars) can be seen east of Antofagasta. The aridity of the region has also led to the formation of nitrate deposits. The large bright patch in the center of the scene is an old nitrate strip-mining area.

A linear break in the landscape of the desert can be seen running north to south approximately 20 km (12 mi.) from the coast. This is the Atacama fault zone. Its most recent movements thrust the western coastal ranges up relative to the land to the east. (R.D.F.)

Bolivian Altiplano

Bright points (below) are snow- and ice-covered volcanic peaks on the western edge of the Bolivian Altiplano within the Atacama Desert. Composed of an elevated surface, generally over 4000 m (13,000 ft.), here the Altiplano is studded with volcanic peaks rising up to 6000 m (20,000 ft.). Prevailing northwest winds sweep the surface, scouring grooves in the landscape and carrying off great quantities of dust to the southeast, which are then deposited on the plains of Argentina. A dust storm sweeps across a low-lying area in the southeastern corner of the scene. The area may be the site of a gigantic caldera, where in the geologic past ash and other debris were erupted and spread over the land.

In the lower center of the scene is a dark smooth basin, the Salar de Atacama, one of the largest salt flats in South America. A crust of sulfate minerals (gypsum and anhydrite) covers the dark center of the basin. The surrounding bright areas to the east are covered for the most part by chlorite salts; seasonal flooding permits fresh white salt to coat the surface. A bright strip in the lower part of the salar is a lithium salt extraction facility. Lithium-laden brines, pumped from underground waters within the salar, evaporate in surface pools.

Calama, a city of 81,000 at the upper left, is the largest oasis in the desert. It supports the mining activities of the province. Just to the north of Calama lies Chuquicamata, one of the world's largest open-pit mines. (R.D.F.)

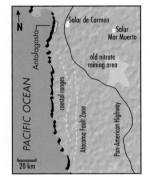

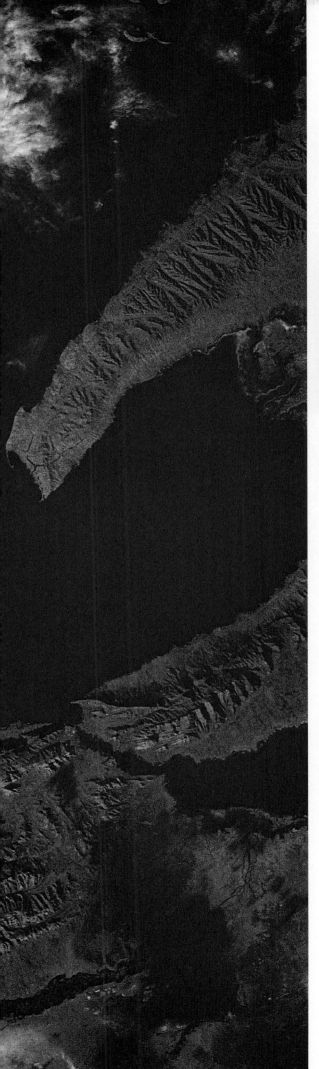

Lake Titicaca

In the central Andes, two roughly parallel cordilleras, the Oriental (eastern) and the Occidental (western), are separated by a high semiarid basin. Lake Titicaca (above), which lies in this basin, is, at 3810 m (12,500 ft.), the highest major navigable lake on Earth. The lake covers about 8300 sq. km (3,240 sq. mi.) and its maximum depth exceeds 275 m (900 ft.).

Most of Lake Titicaca's surplus water is lost to evaporation under the bright sun and dry air. The lake is fed by the melting snows of the Cordillera Oriental. Run-off from the volcanic mountains of the Cordillera Occidental also drains into the lake.

A deeply eroded volcanic cone, Cerro Ccapia, lies near the lake's only outlet, the Rio Desaguadero, which flows south to Lake Poopo, a large, marsh-fringed salt lake. Like many large bodies of water, Lake Titicaca stores heat, which moderates the region's harsh climate and makes agriculture more productive near the lake.

The farmers living along the lake's shores are Aymara speakers, descendents of a culture that was conquered by Quechua-speaking Incas. The Aymara cultivate barley as a commercial crop and quinoa, a protein-rich grain, and potatoes as subsistence crops. These crops are grown on ancient stepped terraces that climb the slopes above the lake.

Colca Canyon

The Río Colca rises on the Altiplano of southern Peru and flows westward, cutting a deep canyon that runs through the Cordillera Occidental and diagonally through this image. The Colca drops from about 4500 m (15,000 ft.) at its sources to the Pacific over a course of 315 km (195 mi.). Over its relatively short, steep run, the river has formed what may be the deepest canyon on Earth—almost 3000 m (9,800 ft.) from the bottom of the canyon to the summit of nearby peaks. A small tarn, or glacial lake, in the shadows just north of Nevado Quehuicha is considered by many to be the source of the Amazon.

In southern Peru the Cordillera Occidental is a chain of volcanoes. In late 1986, Sabancaya, a volcanic cone on a flank of Nevado Ampato, erupted, venting sulfurous gases and discharging ash.

As Sabancaya's eruptions continued during the next several years, ash accumulated on the snow and ice cover of the volcano Hualca Hualca. Snow and ice melting under the ash produced mud flows that ran down into the Colca Valley. In July 1991, earthquakes set off mud slides that buried the village of Maca and damaged several others.

Sabancaya means "Tongue of Fire" in Quechua, and the name suggests that it was active sometime after human beings arrived in the Colca Valley. Lava flows indicate that Sabancaya has erupted since the last glacial period in this region. Post-glacial lava flows can be seen entering glacial valleys to the west of Sabancaya.

The Colca Valley is windswept, steep-sided, chilly, and semiarid. But under the Incas the valley thrived as an agricultural region. Farmers grew maize, potatoes, and other foods in rich volcanic soil that was back-filled into broad terraces on the valley floor and narrow, stepped terraces up the slopes. To water the crops snowmelt was channeled down the slopes to the terraces.

The population of the Colca Valley plummeted after the Spanish conquest, and today nearly half of the terraces remain abandoned.

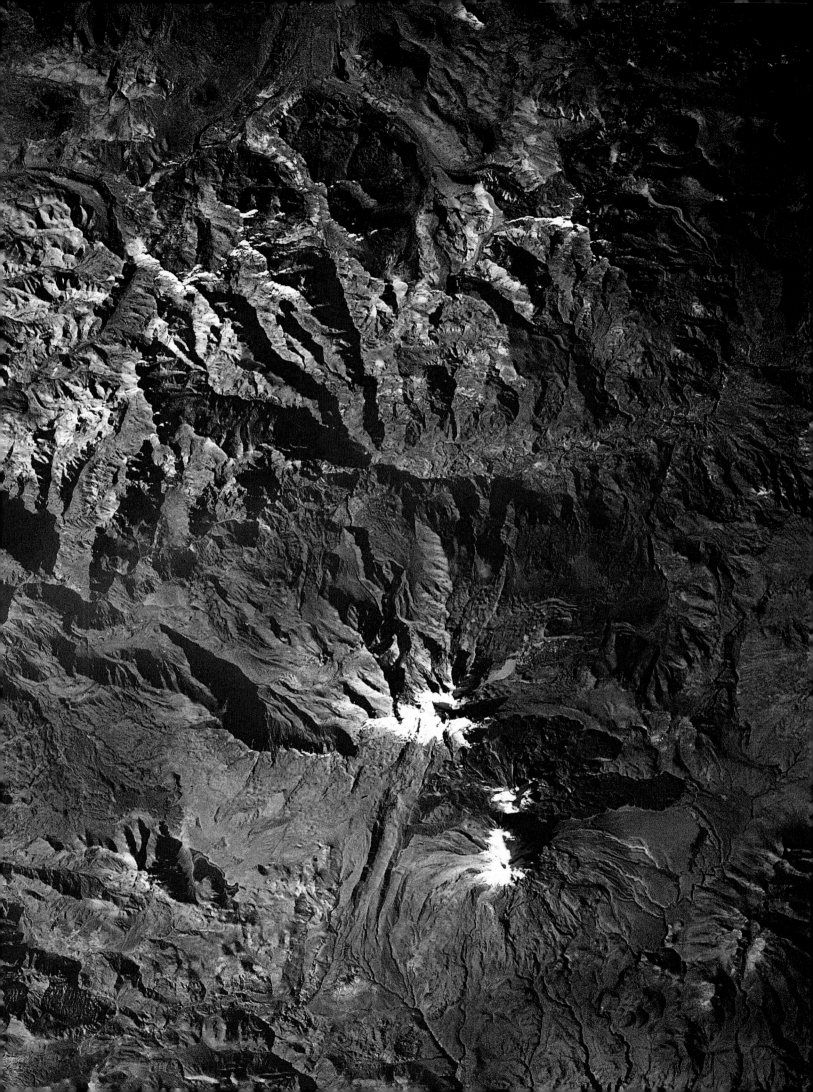

THE LAND OF BOOM & BUST

BEFORE EUROPEANS ARRIVED IN SOUTH AMERICA, ITS CULTURES used the continent's lands in a variety of ways. Hunters and gatherers roamed the Gran Chaco and the pampas. In tropical forests, wandering bands, held together and organized by kinship, practiced shifting farming. On the more fertile river banks of the Amazon, along Peru's coast, and in the eastern highlands of the Andes, chiefdoms established permanent agricultural settlements. About 2,500 years ago the Chavin culture of northern Peru had an agricultural system that could support an urban population. The agricultural state in South America reached its zenith under the Inca empire.

After the Europeans arrived, only in remote regions did traditional land use persist. The Spanish established the latifundia, or large estate, granted by the king. Holdings were passed on to the landowner's eldest son. As in Middle America, the hacienda system developed as a form of latifundia. Haciendas were large tracts farmed or ranched, usually inefficiently. Campesinos, or peasants, worked the hacienda for the use of a small marginal tract. In contrast to the inefficiently worked haciendas, the small campesino tracts, or minifundias, were intensively cultivated.

Unlike haciendas, slave-worked plantations specialized in one specific export crop, such as sugar, bananas, or coffee. Plantations now employ very low-paid laborers. An offshoot of the plantation system was the equally exploitative Amazon rubber-collection industry, which boomed in the 19th Century. The seringueiro, or rubber collector, roamed the tropical forest finding and tapping wild rubber trees and then curing the collected latex on a stick over a fire. The cured rubber was then hauled to a collection point on the river, where an agent decided the value of the rubber. The seringueiro invariably remained in debt to a rubber baron, who lived in luxury and attended the opera in Manaus. The Amazon rubber boom went bust at the turn of the century. But ser-

The Iguazu River can be heard several miles away as it roars over a 4-km-wide (2.5-mi.) crescent on the edge of a basalt plateau between Argentina and Brazil. Water crashing on the rocks below causes a permanent cloud of mist to hover over this, the world's widest waterfall (above). Guarani Indians call it "the place where clouds are born." On both sides of these falls are protected national parks totaling 1490 sq. km (925 sq. mi.).

ingueiros still collect rubber and Brazil nuts. Today many seringueiros belong to a union that advocates the nondestructive use of the Amazon forest. This often puts them in violent conflict with cattle ranchers who want to clear the forest for pasture.

Nonagricultural land use also has a tradition of boom and bust cycles. Tin and copper mining has declined as global demand has diminished. The nitrate boom of the 19th Century ended with the development of synthetic fertilizers. In the 1950s Peru's coastal fishery boomed as larger fishing vessels and new processing plants were established. The industry produced fishmeal and was based mostly on a species of anchovy that thrived in the cold upwelling waters offshore. In 1973 the industry went bust as a result of overfishing and El Niño, a warm ocean current that drives away the anchovies.

The coca-leaf market is slumping as supply far exceeds demand. Alternatives to coca-leaf production are few. Attempts to cultivate citrus fruits in coca-growing regions are thwarted by the lack of transport in the remote Andes of Bolivia and Peru.

In the Andes some campesinos are practicing an old form of subsistence farming. These farmers are producing crops for domestic or community use, and they are planting many different crops, including grains such as maize and quinoa and tubers, which include the potato. Intercropping, or planting several complementary crops in the same plot, reduces a family's vulnerability to crop failure. Maize is planted with lupines, which add nitrogen to the soil, and with squash, which covers the soil and prevents erosion. Chickens and guinea pigs are raised as sources of protein. Larger animals, such as llamas, alpacas, and sheep, are kept as a source of wool. Diversification of production reduces the disasters produced by the failure of a single crop or disease running through an animal herd. On a continental scale, diversification could break the boom-bust cycle and possibly reduce pressure on the land and its resources.

Cataratas del Iguazú

In this image (above) along the Brazil-Argentina border, a sharply angled dark swath marks national parks that were established by the two nations to protect some of the original forest from land-clearing and logging. The parks encompass a long, sinuous stretch of Río Iguazú, which separates the northern end of the Argentine province of Misiones from the Brazilian state of Paraná.

Upstream of its confluence with the Paraná River, the Iguazú drops 82 m (270 ft.) through an array of waterfalls and rapids separated by many small wooded islands. This natural wonder, the Cataratas (Falls) del Iguazú is horse-shoe-shaped and 4 km (2.5 mi.) wide. The Iguazú is funneled into a narrow canyon after its misty plunge down the falls.

To the north of the Paraná-Iguazú confluence, near the city of Foz do Iguaçu, is the massive Itaipú Dam, a joint project of Brazil and Paraguay. The dam is 10 km (6 mi.) long and 61 m (200 ft.) high. Itaipú's turbines have an output of about 12,600 megawatts of power, making it the world's largest hydroelectric project.

Bolivian Lowlands

An image (below) of Bolivia's eastern lowlands distinctly shows the transition from dense, humid jungle (bright red), to less dense, xerophytic (dry-loving) vegetation, called chaparral, in the south (light blue).

There is also a transition in landforms, from the dissected plains of the Brazilian Shield in the upper right, to the low-lying wetlands of the northern Chaco in the south. The demarcation occurs at a scarp run-ning diagonally across the upper portion of the image. The image, acquired during the October-May wet season, shows high water levels in the Rio Grande and in Laguna (Lake) Concepción.

Varying agricultural pat-terns are evident in the image. Near the bend of the Río Grande, some of the nearly square fields are tended by Japanese colonists. The rest are large cattle farms with mixed crop lands.

Near the southwest corner of the image, and again near the center of the image, are seen the linear fields of Men-nonite farming communities. The Mennonites, a religious sect that originated in 16th–Century Europe, emigrated to South America in the 20th Century.

Bolivian campesinos (peas-ant farmers) farm small, linear plots along the roadways east of the river. Most notable are the Bolivian national farms, where central villages are sur-rounded by radiating farm fields about 3 km (2 mi.) wide. In the upper right-hand corner, indigenous farmers tend extremely small plots (pink color) along the savan-nas of the Brazilian Shield.

Land-clearing for farming is rapidly expanding. Evidence can be seen by comparing this image with the one on the opposite page. A farm settlement in the southeastern corner of this image did not exist in the image on page 233, acquired only seven months earlier. (T.B.J.)

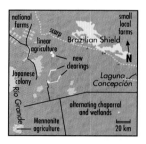

San José de Chiquitos

This dry-season image (above), covering an area near the scene on page 232, reflects seasonal differences. Laguna Concepción has changed shape; forests, deep red on page 232, vary here from orange to dark blue-green. Due to overall dryness and leaf loss in the blue-green areas, the dry season is signaled here by the brightness of the savannas at the upper left and the pink palm wetlands surrounding Laguna Concepción. Standing water made the wetlands nearly black in the previous scene.

Xerophytic (dry-loving) chaparral dominates the lower left corner, except for the bright-red water courses. The town of San José de Chiquitos lies below the center of the image, along the east-west road and railroad from Santa Cruz to the Brazilian frontier. The railroad crosses the Serranía de San José, a mountain range soaring almost 150 m (about 500 ft.) straight up from the level chaparral. Several plateaus and sharp scarps at the lower right have vegetative cover ranging from grasses alone (light blue) to a mixture of grass and xerophytic shrubs (pink to red). Limestone, with sinkholes evident, lies just to the northeast of San José.

Wetlands in bright red demarcate the edge of the Brazilian Shield, which covers the upper two-thirds of the scene.

Just below the center lies a large Mennonite agricultural development. Smaller, more traditional farms are near San José and along the savannas at the upper left. Because large-field farming here speeds up erosion, farmers have begun to leave rows of trees that form cortinas, or windbreaks, which reduce wind erosion of the soil. (T.B.J.)

(previous pages)
Lands Wet and Dry

Wet and dry meet in this image that shows a huge wetland, the Pantanal, bordering the largest semiarid region on the continent, the Gran Chaco. Here the border between Bolivia and Brazil approximates the boundary between near-desert and wetland.

The Gran Chaco extends from northern Argentina, across western Paraguay, and into southeastern Bolivia. Thorny desert scrub and the tough quebracho tree cover much of the Chaco. The word "chaco" is Quechua for "hunting lands." Fierce Indians and a harsh climate deterred early Spanish colonists from settling here. In the 19th Century, a few European settlers moved cattle into the Chaco and established remote estancias, or ranches.

The 140000 sq. km (55,000 sq. mi.) Pantanal is the Earth's largest tropical wetland. This image was acquired at the end of the Pantanal's summer flood season. On the fringe are pools filled by the summer floods. Darker patches indicate wet soil, which, in the dry winter, will become rich grasslands.

Mouth of the Amazon

A great sweep of sediment flows past sandy islands at the mouth of the Amazon on the Atlantic coast of Brazil (right). One island, Marajó, is more than 250 km (155 mi.) across. Near the mouth numerous channels, mudflats, and other islands are built up by the sediments that the river carries. Drawing from the largest drainage basin of any river in the world, the Amazon carries enormous amounts of sediments eroded along its course.

Northward-flowing ocean currents, subsidence (the sinking of sediment and sea floor off the coast), and a strong Atlantic tidal bore (called pororoca) all combine to inhibit the forming of an extensive Amazon delta. The muddy color of Amazon sediment extends well out into the Atlantic. The fresh waters of the river mingle with the salty ocean as far as 160 km (100 mi.) from shore, which is an indication of the immense volume of Amazon waters. More water flows from the Amazon than from the combined flow of the next eight largest rivers on Earth. Estimates of the volume of water carried by this mighty river reach as high as 25 percent of all the world's river water.

Floods frequently wash over many areas on the Amazon's long course. A large part of Marajó Island is inundated

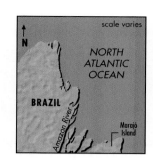

each rainy season. Grasslands on the east side of the island support cattle and water buffalo. Across a waterway southeast of Marajó Island is the city of Belém (not shown in image), the Amazon's main port for ocean-going vessels. In the 17th Century sugar cane was cultivated in the Belém region. Rice, cotton, and coffee later replaced sugar as cash crops. But by the mid—19th Century the agricultural importance of the area had declined and the economy was based on Belém's role as a thriving center of trade.

The Amazon is very deep at its mouth (about 90 m or 290 ft.) allowing ocean-going ships to sail upriver. The river is so navigable that large vessels can travel inland for some 1600 km (1,000 mi.) to the city of Manaus. Small ships can sail as far as Iquitos (in Peru) on the western side of the continent—Peru's "port on the Atlantic." The Amazon has its source in the Andes only about 160 km (100 mi.) from the Pacific coast. It travels nearly 6400 km (4,000 mi.) eastward across the continent and is joined by approximately 1,100 tributaries.

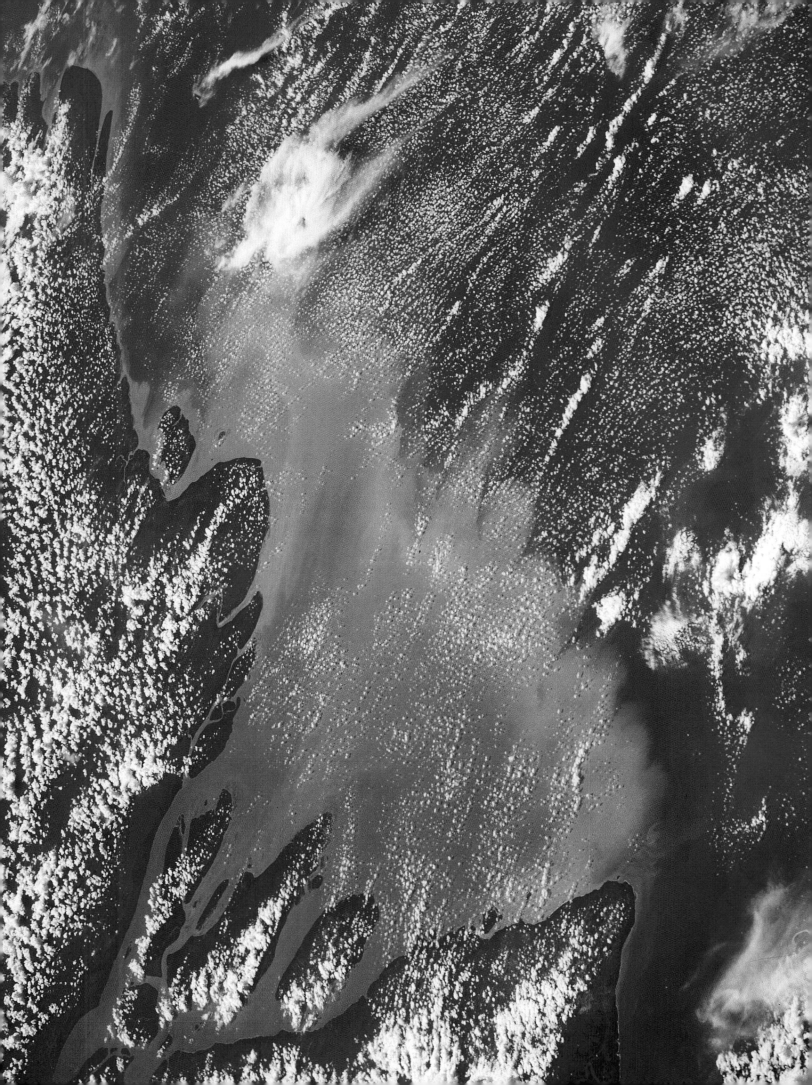

Nazca, Peru

The bright red signatures of the Rio Nazca and the Rio Ingenio gorges (above) are separated by a gray desert surface. Mineral oxidation produces the desert's dark-gray varnish. Lying just beneath the dark patina is a lighter layer of coarse sand.

Along Peru's coastal desert, or costa, a series of platforms, cut by river channels, descend from the western slopes of the Andes. Farmers grow cotton, corn, grapes, and barley along the flat-bottomed river gorges.

Because many of the costa's rivers only flow for a few days a year, farmers have been irrigating with ground-water in the Rio Nazca valley since at least 1,500 years ago. At that time, valley dwellers began digging horizontal tunnels, or puquios, to tap the region's groundwater. Vertical shafts to the stone-lined puquios gave access for cleaning and repair.

The people who built the puquios in the Nazca valley belonged to a thriving agricultural society archaeologists call the Nazca culture. The Nazca are best known for the "Nazca Lines" etched on the land north of the Rio Nazca. The geometric (left), animal, and plant figures were created by removing desert varnish to expose the lighter soil beneath. The Nazca produced the world's largest and most varied collection of geoglyphs, or ground drawings. Geoglyphs are also found in Bolivia and northern Chile, as well as in North America and Europe.

The Nazca geoglyphs include a hummingbird, a monkey, a condor, and other animals, all less than 100 m (330 ft.) long. The most intriguing geoglyphs are the more than 1000 km (620 mi.) of lines that crisscross the desert.

Some of the lines extend for several kilometers and are visible with space-based sensors. The two brightest lines that form a cross (above) are a road and a power line right-of-way. The less bright lines that intersect and cross the modern lines are linear geoglyphs.

The original purpose for creating these linear geoglyphs is unknown. Theories include astronomical calendars and ritual pathways to sacred landmarks.

Río Paraná

Of South America's rivers, Río Paraná is second in length only to the Amazon. It stretches for 4000 km (2,480 mi.) from its headwaters in southern Brazil, along the border between Paraguay and Argentina where it is joined by the Paraguay River, and south through Argentina to the Río de la Plata. The river's extensive delta begins developing some 320 km (200 mi.) from its mouth and is 18 to 64 km (11 to 40 mi.) wide.

This SPOT image shows a region near the northern edge of the delta, east of the river port city of Rosario. Here meandering distributaries (channels flowing away from the main branch of the river) make their way through heavily vegetated swamps. Dark spots indicate pools of stagnant water in ponds and abandoned meanders. Geometric patterns at the upper right indicate farms and citrus groves on the edge of the Entre Ríos Plateau. The name Entre Ríos (Between the Rivers) comes from its location between the lower reaches of the Paraná and Uruguay waterways.

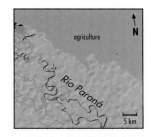

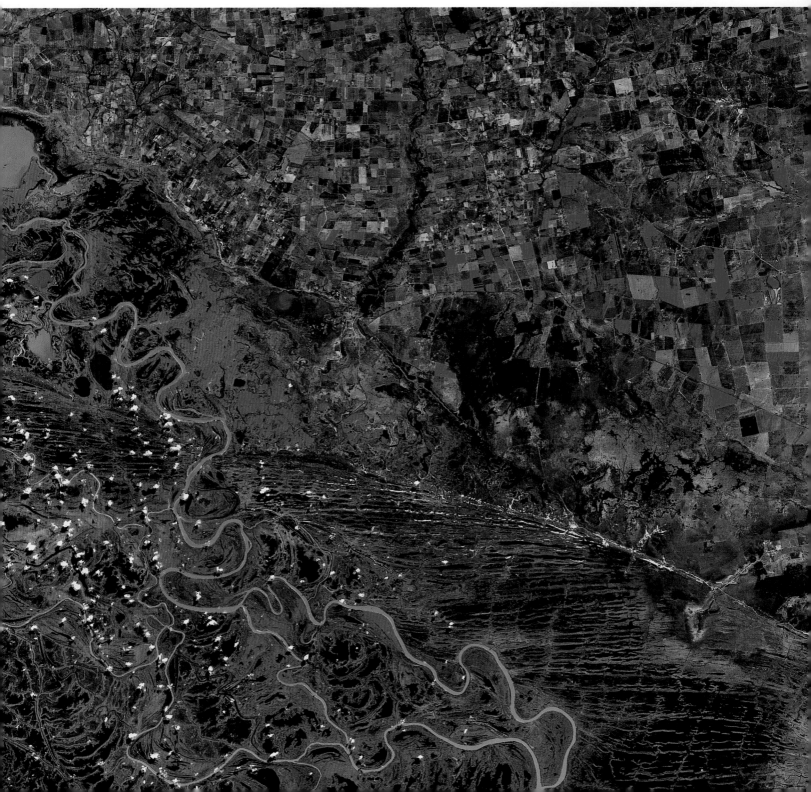

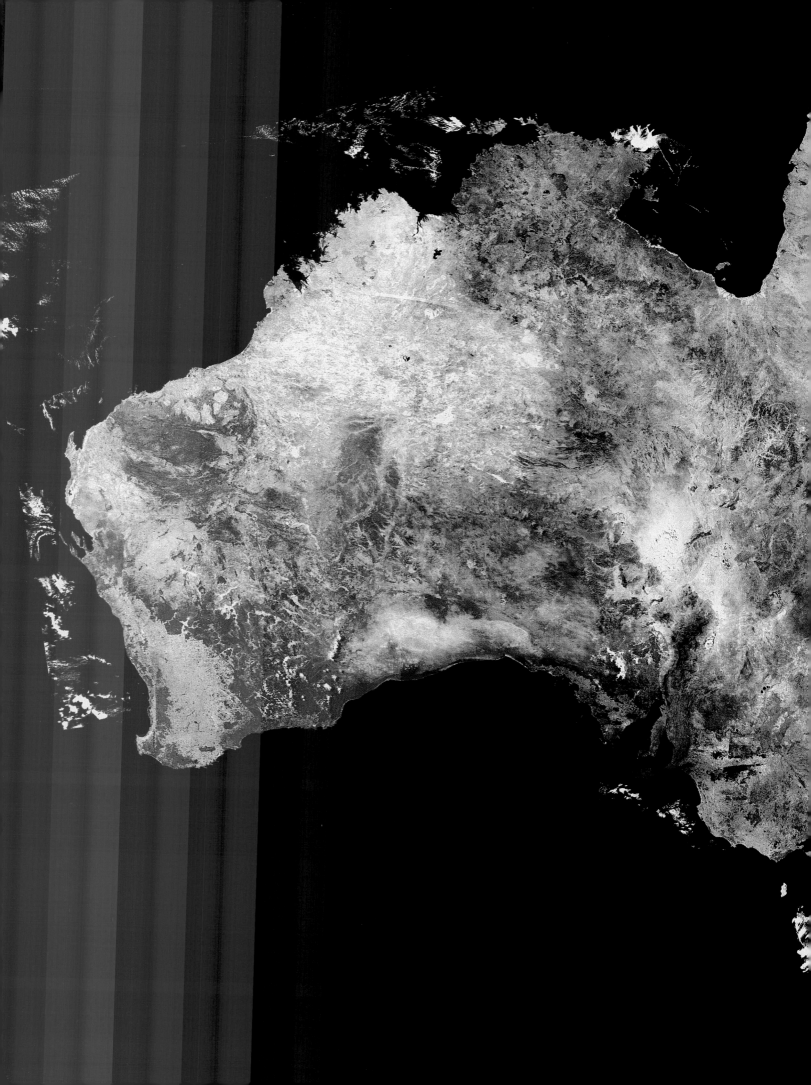

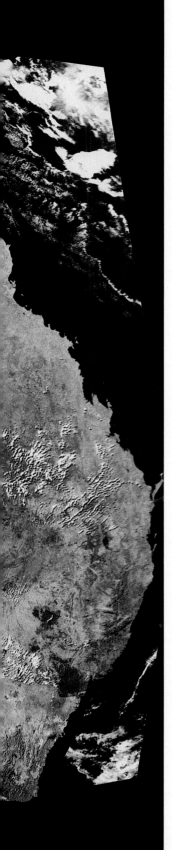

REALM OF THE PACIFIC

THE PACIFIC OCEAN BASIN OCCUPIES ONE-THIRD OF THE EARTH'S SURFACE, an area of about 166 million sq. km (64 million sq. mi.) that stretches from its narrowest and northernmost point, the Bering Strait between Alaska and Siberia, to Antarctica's coast. The Pacific's waters wash the shores of five continents and have the deepest sounding of the world's oceans—11033 m (36,198 ft.) in the Mariana Trench.

Oceania is a huge region that contains the islands of the central, southern, and western Pacific Ocean, including New Zealand and the continent of Australia. This broad definition includes more than 10,000 islands and a range of diverse cultures, such as Melanesian, Micronesian, and Polynesian, along with a European and Asian presence. Although the extent of Oceania is great, most of the area is water. The region's total land area is less than Antarctica's.

About 300 million years ago a large landmass, known today as Gondwana, lay in the Southern Hemisphere. Gondwana consisted of pieces of light continental crust that had broken up and begun to drift apart more than 100 million years ago. The breakup of Gondwana formed the continents of Africa, South America, Antarctica, Australia, and the subcontinent of India.

Earth's smallest continent, the Australian landmass (8 million sq. km or 3 million sq. mi.), is surrounded by the South Pacific and Indian oceans. There are three major physical regions in Australia: the western plateau or shield, the central lowlands, and the eastern highlands. The highlands and the plateau appear darker in the satellite mosaic (opposite page), and some vegetation (yellow) is visible in these regions. The central lowlands are much lighter, with the exception of the thin bands of the Macdonnell and Petermann ranges at the continent's geographic center. Ayer's Rock, one of Australia's best known natural features, is situated between these two ranges.

The terms plateau, lowlands, and highlands are relative, for in absolute terms Australia has the least range of relief of all the continents. The continent sits on stable crust that has not undergone the great collisions that thrust mountains upward in other continents.

Mount Kosciusko in the Snowy Mountains has an elevation of 2230 m (7,320 ft.), making it the continent's highest point. The Snowy Mountains are part of a long chain of mountains, the Great Dividing Range, that parallels the east coast. This range is a major part of the eastern highlands that extend from Cape York to the Bass Strait, separating Australia from the island of Tasmania, one of Australia's six states. Formed by an extension of the eastern highlands, Tasmania's steep-sided central plateau was glaciated during the Pleistocene. The uplift of

AVHRR MOSAIC, REMOTE SENSING APPLICATION CENTRE, DEPARTMENT OF LAND ADMINISTRATION, PERTH, WESTERN AUSTRALIA

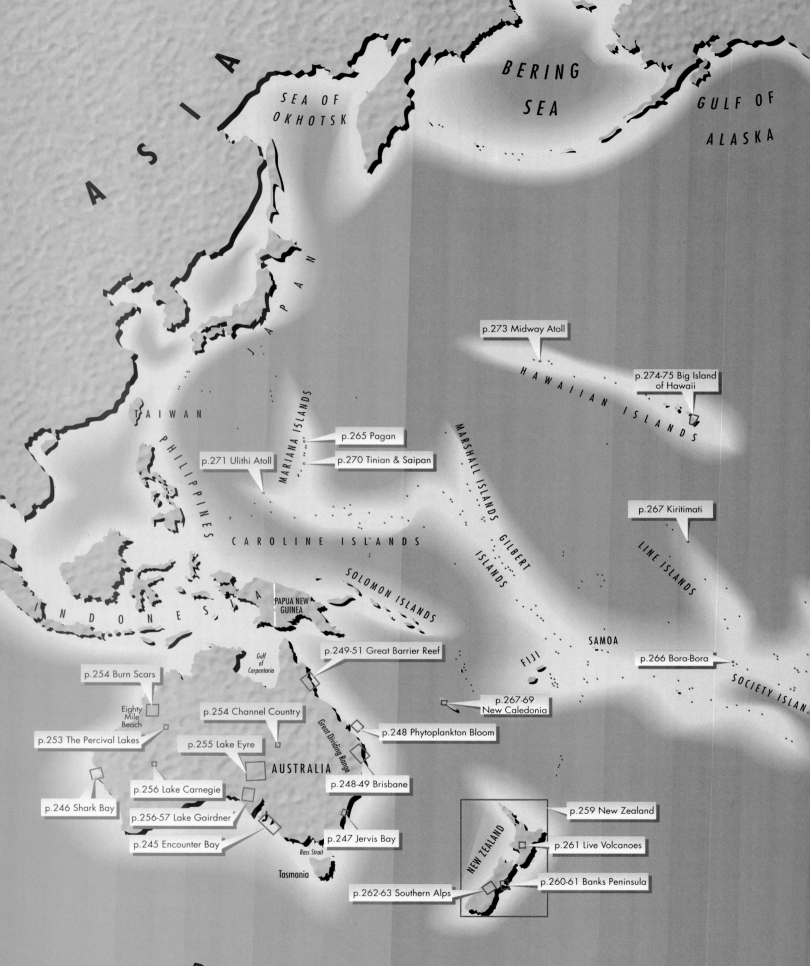

ASIA

SEA OF OKHOTSK

BERING SEA

GULF OF ALASKA

JAPAN

TAIWAN

PHILIPPINES

p.273 Midway Atoll

HAWAIIAN ISLANDS

p.274-75 Big Island of Hawaii

MARIANA ISLANDS

p.265 Pagan

p.271 Ulithi Atoll

p.270 Tinian & Saipan

MARSHALL ISLANDS

p.267 Kiritimati

INDONESIA

CAROLINE ISLANDS

GILBERT ISLANDS

LINE ISLANDS

SOLOMON ISLANDS

PAPUA NEW GUINEA

p.249-51 Great Barrier Reef

SAMOA

p.266 Bora-Bora

Gulf of Carpentaria

FIJI

SOCIETY ISLANDS

p.254 Burn Scars

p.254 Channel Country

Great Dividing Range

p.248 Phytoplankton Bloom

p.267-69 New Caledonia

Eighty Mile Beach

p.253 The Percival Lakes

p.255 Lake Eyre

AUSTRALIA

p.248-49 Brisbane

p.246 Shark Bay

p.256 Lake Carnegie

p.256-57 Lake Gairdner

p.245 Encounter Bay

p.259 New Zealand

NEW ZEALAND

p.261 Live Volcanoes

p.247 Jervis Bay

Bass Strait

p.260-61 Banks Peninsula

Tasmania

p.262-63 Southern Alps

PACIFIC OCEAN

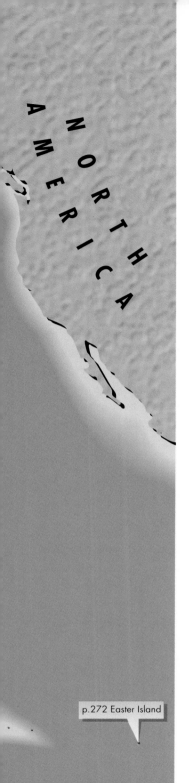

Australia's eastern highlands probably began between 225 and 65 million years ago and included some volcanic activity that may have continued until a few thousand years ago. The steep eastern slopes of the highlands are deeply cut by rivers, such as the Shoalhaven River in New South Wales, while the highlands' western slopes descend gradually toward the center of the continent.

The central lowlands of Australia extend from the Gulf of Carpentaria in the north to the Great Australian Bight in the south. The lowlands include several basins, including the Murray, the Eyre, and the Carpentaria. At the northern end of the lowlands the Carpentaria Basin is partially submerged beneath the Gulf of Carpentaria. Lake Eyre (see page 255) is a large playa, or salt pan, lying at the bottom of a large drainage basin. The salt crusts of Lake Eyre and nearby playas reflect brightly near the center of the continent in the satellite mosaic. The Murray River drains the western slope of the southern Great Dividing Range and is Australia's longest perennial river.

The state of Western Australia sits on an old, fractured continental shield that is made up of a number of blocks and basins and a complex assortment of volcanic, metamorphic, and sedimentary rocks. The shield extends into both the Northern Territory and the western part of South Australia. Broad, flat Nullarbor Plain is formed by thick limestone deposits in South Australia and Western Australia and is bounded on the south by steep cliffs stretching along the shores of the Great Australian Bight.

Australia's climate, with a few local exceptions, can be described in one word: dry. The continental mosaic reveals a landmass with little plant cover and bright, reflective desert regions. Much of Australia lies in the subtropics, a region of warm, descending dry air that produces little rain. Moreover, without mountains and high plateaus, there is little upward deflection of moist air blown in from the surrounding seas, which further reduces cloud formation. The lack of both rainfall and young, high mountains leaves the continent with relatively low rates of weathering. As a result, with a few notable exceptions (see page 244), much of Australia's soil is very old and lacks mineral and organic materials.

Australia has about 26000 km (16,000 mi.) of coastline, whose forms include long sandy beaches, coral and algal reefs, mangrove swamps, and high, steep cliffs. Isolated Australia is, in effect, a large island where evolution took a separate course and produced a number of unique animal and plant genera and species. Australia's fauna includes the egg-laying mammals—the duck-billed platypus and the spiny anteater—along with other mammals such as the kangaroo, wallaby, and koala. These animals thrived in their isolation, particularly the marsupials, which had not endured on other continents where predators were more common.

The first human beings arrived from Asia between 40,000 and 10,000 years ago when sea levels were lower and land bridges connected many of the major islands northwest of Australia. A sea crossing was still necessary between Australia and Asia, but the lower sea levels made the trips shorter. Australia's aborigines have a culture that is about 40,000 years old. This culture developed in isolation, lacked a written language, and was technologically poor. Nonetheless, the aborigines created a culture that is rich in art and mythology. Australia's unusual biological history continually stimulates scholarly interest.

A Coast for all Reasons

FROM BARREN COASTAL DESERTS TO LUSH OFFSHORE REEFS teeming with life, Australian coasts exhibit great variety. There are shipping centers for Australia's vast exports of mineral products, wool, and beef. There are commercial resources, such as zircon that is mined from beach sands. A thriving tourist industry is also sustained by the coast.

Dutch merchant ships carried out much of the early exploration of Australia's west coast in the 17th Century while en route to the colonies in the Dutch East Indies. The outward voyage from the Netherlands was shortened by several months when the Dutch began crossing the Indian Ocean south of 40° latitude, a region known as the Roaring Forties, where the westerly winds were strong and constant. After making landfall on Australia's western coast, the ships turned north toward Java and completed their outward passage.

The British explorer Matthew Flinders, who first sailed to Australia in 1795, circumnavigated this island continent from 1801 to 1803. His detailed charts of a large proportion of the Australian coastline were important to the charting of the continent.

Australia's coast has many large bays and gulfs. (See opposite page and pages 246-47.) In the north, the shores of the rectangular-shaped Gulf of Carpentaria are lined with old beach ridges and marshy lowlands. Exploitation of large resources of bauxite on the Gulf's western islands and manganese along the northwestern and eastern shores began in the 1960s. In the northwest, the broad arc of Eighty Mile Beach marks a low-lying region with sandy shores, some dunes, and marshes. The south coast is dominated by the vast bay called the Great Australian Bight, known for its choppy waters. The Bight is bordered in the east by low cliffs at the edge of the vast plateau of the Nullarbor Plain. In the west a narrow coastal lowland marks the edge of the plain.

In the northeast is Australia's most striking offshore feature, the magnificent Great Barrier Reef, the longest coral reef structure in the world. (See pages 250-51.) It is a series of reefs stretching for about 2000 km (1,200 mi.) parallel to Queensland's coast. Barrier reefs form along coastlines where currents bring the warm waters necessary for coral's optimal growth. Corals are simply structured marine animals that generally live in colonies. Some varieties, called hard corals, secrete protective limestone coverings. The reefs are built up on these "skeletons" that dead corals leave behind. Detritus from coral remains, which has been worked and reworked by erosive waves, also fills in the reefs.

Coral reefs not only need warm water but also clear water to thrive. Sunlight must penetrate the water, enabling microscopic plants in the coral's tissues to photosynthesize and produce oxygen that is utilized by the corals. Too much sediment in the water blocks the sun and clogs the reefs, interfering with coral feeding and breathing cycles.

An amazing variety of fauna inhabits the barrier reef: corals and worms, lobsters and crabs, tiny, brightly colored fish, and enormous groupers and sharks. The potato cod and the Queensland grouper, indigenous only to Australian waters, can actually reach lengths of 2 m (6.6 ft.) The world-renowned giant clams live in great numbers here. These reach about 1 m (3.3 ft.) across. Contrary to common myth, they do not eat divers!

The pressures of tourism and oil exploration have threatened the fragile environment of the Great Barrier Reef. But the Australian government has taken steps to protect the reefs from overuse and exploitation by the designation of the Great Barrier Reef Marine Park. Laws prohibit any damaging of the reefs or the touching or removal of any reef animals or materials. Even the type of shoes to be worn when walking on the reef is regulated. These limitations, which are necessary to prevent the abuse of this natural asset, may help sustain the Great Barrier Reef as an international ecological treasure.

The Great Barrier Reef forms a protective breakwater more than 1932 km (1,200 mi.) long for the coast of Queensland, Australia.

Encounter Bay

Encounter Bay curves along the southeast coast of the state of South Australia. Lying to the southeast of the bay is the Younghusband Peninsula, a long sand bar formed as sediments carried along the shore are deposited in calm waters. A quiet lagoon rests behind the bar. Off the coast to the west is Kangaroo Island, the home, as one might guess, of a large number of kangaroos. The Fleurieu Peninsula juts toward Kangaroo Island.

Along the coast north of the peninsula is Australian wine country, which produces much of the wine and brandy consumed in Australia.

Encounter Bay was named by explorer Matthew Flinders, who commanded the ship *Investigator* on an 1801-1803 expedition along Australia's coastlines. On this journey, he encountered a rival explorer, the Frenchman Nicolas Baudin. They exchanged information, and Flinders later named the bay where they met to commemorate the meeting.

Shark Bay

Narrow islands and peninsulas isolate Shark Bay, on the coast of Western Australia, from the Indian Ocean. The irregularly shaped landforms are former coastal sand dunes that have been surrounded by

water as the sea level rose. Circulation of sea water in the bay is poor, with salinity and water temperature increasing toward the southeast end of the bay.

Other factors creating the high salinity in the bay are the low rate of rainfall and the high rate of evaporation in a region with a hot, dry climate. Water flows infrequently from the two intermittent rivers that enter the bay, which also makes the enclosed waters more saline.

The high salinity and water temperature in Shark Bay prevent the growth of coral. But reef-building algae are present. Algal reefs, also called stromatolites (right), are built by mats of blue-green algae that grow in shallow water and produce a calcium carbonate precipitate, known as aragonite. It cements pieces of shell and grains of sand together, forming a sedimentary structure. These structures take a variety of shapes, ranging from club-like hummocks to

broad domes and platforms. The shape of these stromatolites is probably controlled by such environmental agents as currents and tides.

The shores of Hamelin Pool contain club-shaped and domed stromatolites. Reef-building algal mats are present in Freycinet Estuary.

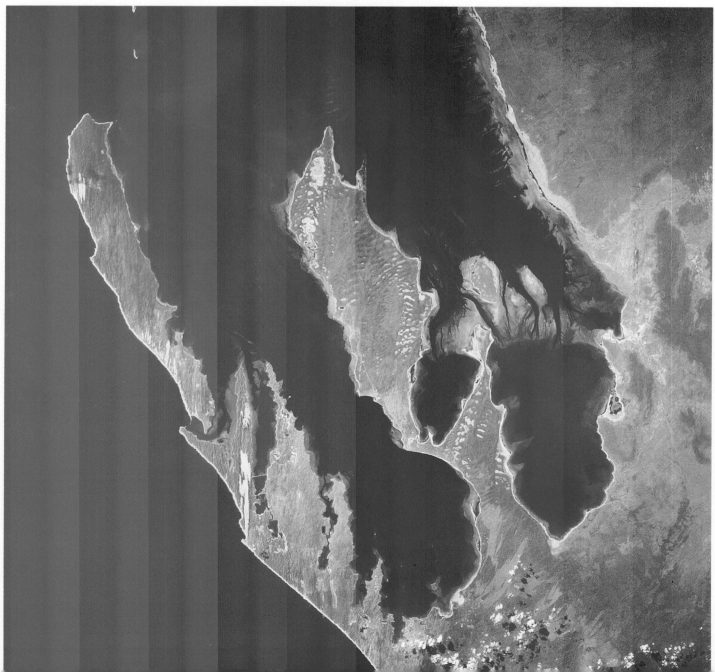

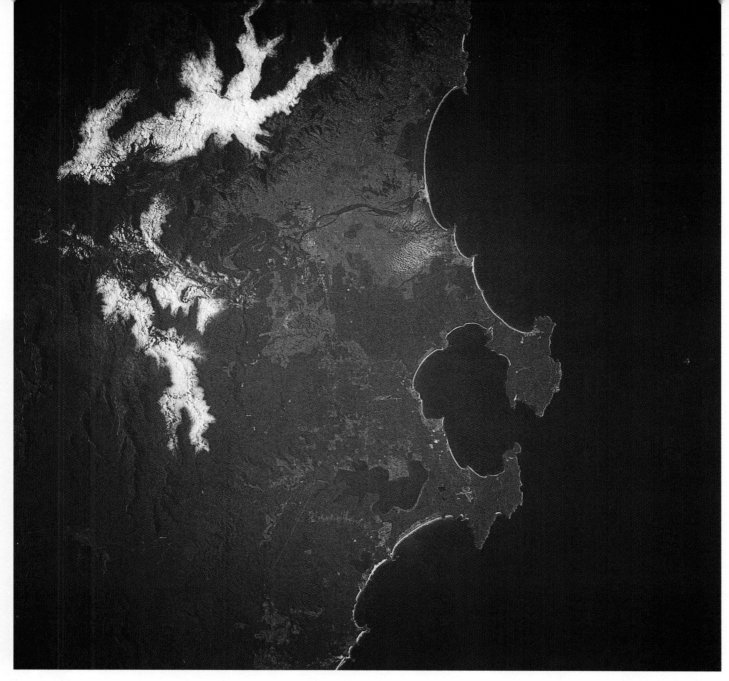

Jervis Bay

Ragged banks of fog hover over the valleys along the south coast of New South Wales, a region of forests, dairy farms, and fishing ports. This Space Shuttle photograph was taken shortly after sunrise in October during the Australian spring. On clear, cool nights the land cools as heat energy accumulated during the day radiates skyward. As the ground radiates heat and cools, the air above also cools. The cool, moist air has collected and condensed in the valleys, forming the fog visible in this photograph. As the sun rises higher during the day, the ground and the air above it will continue warming. The water droplets forming the fog will change to water vapor. The radiation fog will appear to lift or burn off.

Most people of the region live in coastal towns, such as Nowra, near the mouth of the Shoalhaven River. Cattle raising is the predominant mode of agriculture, and the climate sustains grazing for sizeable herds. Pasture land is visible as the lighter areas on either side of the Shoalhaven. To the south of Nowra is Jervis Bay, site of a fine harbor that is about equidistant from Sydney and Canberra. Jervis Bay is technically an embayment formed by two arms, or tombolos. They usually occur along submerging coastlines; sea level on the coast of New South Wales has been rising since the end of the Ice Age, or Pleistocene epoch.

Offshore islands near subsiding coasts are often remnants of mainland hills or mountains. The seaward shores of these islands often have steep cliffs created by erosion caused by waves. In the landward direction, wave energy is diminished and eroded material is deposited, forming a bar or spit that connects the former island with the mainland.

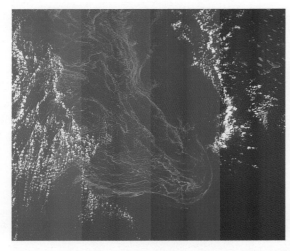

Brisbane

The capital of the state of Queensland and an important port, Brisbane (below) lies on the Brisbane River about 23 km (14 mi.) inland from Moreton Bay. The city has its origins in a penal colony for hardened criminals sent there from Sydney. By 1831 more than 1,000 prisoners were held here. Until 1842 free persons were prohibited from settling within 80 km (50 mi.) of the prison complex. Although transporting convicts from England to the Australian mainland was abolished in 1840, settlement of convicts as laborers resumed in the Brisbane area and continued until 1850.

Brisbane was named for Sir Thomas Brisbane, governor of New South Wales and a patron of science. He built an astronomical observatory near Sydney to increase knowledge of the stars of the Southern Hemisphere.

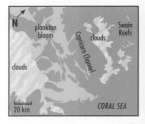

Phytoplankton

Tendrils of green disclose a large bloom of phytoplankton, a microscopic drifting marine plant, off the coast of Queensland (top). The bloom, 150 km (93 mi.) long, extends through the Capricorn Channel. Swains Reef, the largest group of reefs at the southern end of the Great Barrier Reef, is visible on the north side of the channel.

The algae here typically grow as thin filaments that coalesce and form small brown flecks about 2 mm (0.08 in.) in diameter. When the floating mass of algae rots, bacteria in the seawater near the bloom increase dramatically. The bacteria then consume much of the dissolved oxygen.

Marine animals around a plankton bloom may die from the toxins that some algae species produce or from lack of oxygen.

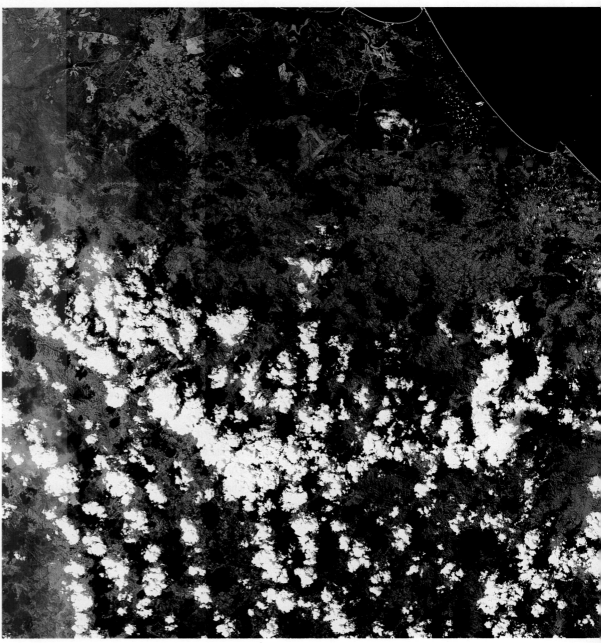

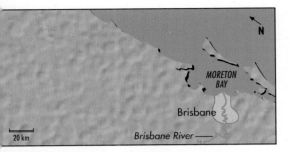

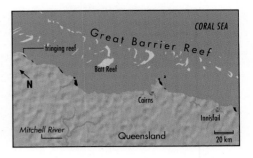

(following pages, map above)

Great Barrier Reef

Several reef forms, each shaped by the coastal environment, appear in this image of the Great Barrier Reef. Fringing reefs (far left) lie in shallow water. Ribbon reefs, marking where the continental shelf abruptly drops into much deeper water, are visible to the north of Cairns. Behind the outer barrier of ribbon reefs lies a group of curving platform reefs. During a relatively short time—9,000 years of building—about 20 m (66 ft.) of biologically produced coral has accumulated.

South of Cairns, the continental shelf begins to widen, and platform reefs predominate. Exposed to the swells that roll in from the Coral Sea, platform reefs accumulate sand on the leeward side and become crescent-shaped.

Living reefs are fragile ecosystems, and Queensland's increasing coastal population threatens them. In 1975, a preservation law established the Great Barrier Reef Marine Park, banning mining and drilling, to protect the reef.

A CONTINENT OF DESERTS

ABOUT 75 PERCENT OF AUSTRALIA CONSISTS OF ARID AND semiarid land. This is the largest proportion of desert lands found on any continent. But Australian deserts are not like the great, perennially dry deserts of northern Africa and the Middle East: hyperaridity and active dune processes are lacking. Rains do fall intermittently on the Australian desert lands, and summer monsoons can bring irregular rainfall to the braid of dry riverbeds called Channel Country (see page 254). Most Australian deserts have 30 to 60 percent of their surfaces covered by vegetation. Although large dunes occur over vast areas, most dunes are stabilized by desert grasses and shrubs.

Several deserts have been delineated on the barren lands of central and western Australia. There is the Great Sandy Desert of northwest Western Australia (opposite page); the Great Victoria Desert, which crosses from Western Australia to South Australia; and the Simpson Desert (see page 254), located at the juncture of the Northern Territory, Queensland, and South Australia. They are all sandy deserts. The Gibson (see page 256) of central Western Australia and the Sturt of east-central Australia are stony deserts. They have tablelands capped by silica-hardened soil crusts. The tablelands rise amidst plains blanketed with pebbles known as gibbers.

The sandy deserts often exhibit a red coloration, which is due to the weathering of iron oxide coatings on individual sand grains. The hue of the sands explains why Australia is often referred to as the "Red Continent." The sandy deserts are also characterized by broad expanses of strikingly parallel linear dunes. These sand ridges average around 10-15 m (30-50 ft.) in height and can extend uninterrupted for as far as 160 km (100 mi.). They are longitudinal dunes, that is, dunes that are oriented parallel to the prevailing wind direction. The orientations of the sand ridges mimic the major Australian wind patterns across the continent. These dunes are especially well-de-

veloped in the Great Sandy and Simpson deserts and are somewhat less prominent in the Great Victoria Desert.

The numerous, closely spaced sand ridges made early journeys through these regions especially arduous for explorers who tried to cross them in the late 19th and early 20th centuries. They found a place where, in the words of poet Henry Lawson, "the scrubs and plains are wide, With seldom a track that a man can trust, or a mountain peak to guide. . . ." To this day, much of the terrain has not been documented in detail from the ground. Aerial and orbital photography, however, is ideal for tracing the extent and orientation of the large linear structures and for mapping the barren and sparsely populated arid zones.

Desert salt flats are common throughout the Australian interior. Most are small and rarely flood. Evaporation often leaves mineral deposits in unusual formations.

A notable feature of the Australian deserts is their abundance of playa lakes (opposite page and pages 255-56). These playas, also called salt pans or salt lakes, are usually dry, but they can fill with water when scarce rains activate ephemeral streams or cause flash floods. Without a regular supply of water the lakes evaporate, leaving behind salt and gypsum deposits in fascinating patterns. These remnants of what were once truly lakes are testament to more humid climates that existed ages ago in previous geologic epochs. The salt pans of the Great Sandy Desert are sometimes arranged in linear patterns that trace the courses of ancient drainage systems that once brought a continuing supply of water to a land rich in vegetation.

Australian deserts do not constitute an attractive environment for human habitation—settlements are rare here because of the hardships involved. In earlier days, the deserts created obstacles to the growth of industry by obstructing the shipment of goods and by discouraging commercial development. The mineral wealth of the region, however, has provided incentives to develop the land. Mining towns have sprung up. But water and supplies must be hauled in from great distances to support modern life in this barren land.

The Percival Lakes

A string of playas, or salt pans, forms the Percival Lakes, part of a chain of such lakes extending more than 400 km (250 mi.) across the Great Sandy Desert. Stable longitudinal dunes can be seen bordering the Percival Lake system in this view of its eastern end (above). The alignment of the salt pans suggests they are the remains of a river course that flowed when the region experienced a more humid climate millions of years ago.

Aligned lakes such as these are common in Western Australia. They may mark the sites of aquifers, underground water sources, that would be vital to settlement and industry in the region. The lakes, which were not documented until 1934, were named for Edgar W. Percival, a famous Australian pilot, designer, and manufacturer of aircraft.

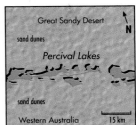

Great Sandy Desert

N

sand dunes

Percival Lakes

sand dunes

Western Australia   15 km

Burn Scars

Repeated burning of desert grasses produces these ground patterns, a common sight in arid regions of Australia. The lightest colored areas, bright because reflective desert sand is uncovered, mark recent fires. Darker areas are places where grass has grown back, as along creek beds in the photo below.

The fires are caused by lightning, by intentional burning to stimulate new grass growth for grazing, or by hunters. Comparison of images over several years can indicate the time it takes for land and grasses to regenerate.

Channel Country

Most Australian rivers flow only intermittently, responding to seasonal rains. Dry channels of the Diamantina River (right) form a terrain of densely intertwined channels called Channel Country.

Australia's eastern mountain ranges block moisture-carrying trade winds and thus keep the continent's interior and river channels dry throughout much of the year. Summer monsoons, however, can bring heavy rains that overload the sediment-clogged channels, causing short-lived floods. In severe floods, the streams coalesce into broad sheets of water up to 80 km (50 mi.) wide.

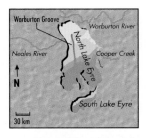

Lake Eyre

Lake Eyre, in the driest part of Australia, is a salina, a salt-encrusted lake bed rarely covered with water. Dry river channels enter the Lake Eyre basin from the northeast. The rivers, which rise in the uplands in eastern Australia, often receive enough precipitation to begin flowing toward Lake Eyre. But the river water is lost to evaporation and seepage in the more arid regions downstream. Only after unusually heavy rains does the lake fill.

This photograph (below) was taken in November 1984, a year with above average rainfall. Water has filled most of South Lake Eyre and the southern half of North Lake Eyre. The Warburton Groove, which crosses the salt flats, is a channel that conducts the discharge of the Warburton River to the deeper southern half of the lake.

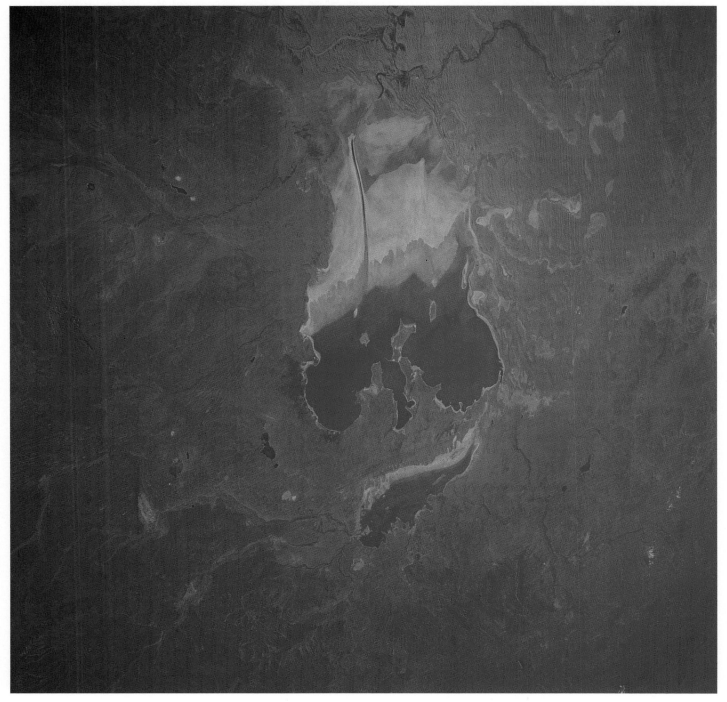

Lake Carnegie

Strange patterns mark Lake Carnegie (right) on the western edge of the immense Gibson Desert (above) of Western Australia. The lake's shallow basin is usually dry, but it holds water after heavy rains, which come infrequently.

Erosion caused by wind has contributed to the intriguing form of the lake deposits, which surround the low rises. Salty clay and mud coat the basin floor, and dunes lie around its edges.

"In Australia alone," wrote a 19th-Century journalist, "is to be found the Grotesque, the Weird, the strange scribblings of nature learning how to write the subtle charms of this fantastic land. . . ."

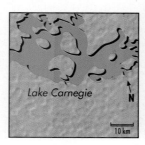

Lake Carnegie

Lake Gairdner

Lake Gairdner (opposite page) is the largest of a group of salt lakes that lie near the base of the Eyre Peninsula in South Australia. Annual rainfall is less than 25 cm (10 in.) per year, and the region is sparsely vegetated with salt bush and scrub. The interior of South Australia has an arid climate where evaporation far exceeds precipitation.

Much of Lake Gairdner's surface is a thin crust of salt covering layers of mud, silt, and clay. As the lake's waters evaporate, dissolved minerals become more concentrated and the water more briny. Subsequent evaporation deposits the dissolved minerals. The concentration of sodium and chloride in the lake's salts is similar to sea water, suggesting that the ocean may have once covered this part of Australia.

The low, rounded hills of the Gawler Ranges lie south of Lake Gairdner. The ranges are composed of old volcanic rock that dates back about 600 million years. Joints in the volcanic rock, which trend northeast-southwest and northwest-southeast, have been eroded into small valleys with flat, sandy bottoms. After a rain—only occasional in this area—these joints channel surface water into Lake Gairdner's brine.

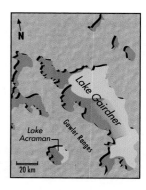

Lake Acraman lies in a circular depression about 20 km (12 mi.) in diameter. A meteorite or comet made the crater possibly 600 million years ago. The eastern shore of Lake Gairdner is partially concentric with Lake Acraman, and there is speculation that the crater may actually have a diameter of 160 km (99 mi.).

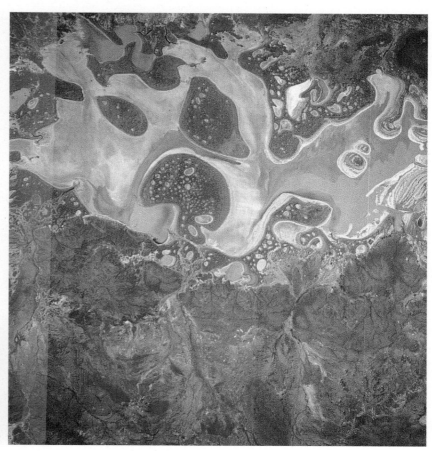

New Zealand—A Land Lifted High

THE COUNTRY OF NEW ZEALAND CONSISTS OF TWO MAIN islands, North Island and South Island, as well as numerous smaller islands and atolls. Among these are Stewart Island, which is just off the south coast and which has a population of fewer than 600, and the Chatham Islands, a volcanic group of inhabited and uninhabited islands to the east.

Although close to Australia, New Zealand has a dramatically different landscape. Its small size, its latitude, and its location on the boundary between two major plates of the Earth's crust have combined to produce a land of cool, wet climate with high lands and young, active geologic structures. In contrast to Australia's stable old weathered surface, New Zealand has active volcanoes in the north and rugged young folded mountains in the south.

The North Island centers on Lake Taupo, an old volcanic caldera, now New Zealand's largest lake. This island is the more heavily populated and is home to the country's largest city and principal port, Auckland. The island's interior is mainly a plateau coated with volcanic ash. A string of volcanoes marks the active zone that bisects the island. (See page 261.) Steaming hot springs and geysers are common. Steam and water fueled by the heat of the volcanoes' magma are important for the production of electricity.

The South Island is very mountainous. The island's major chain, the Southern Alps (see pages 262-63), extends for about 500 km (300 mi.) along the western coast. Repeated periods of mountain glaciation have carved out the rugged topography of the South Island. Both glaciers and glacial lakes are still evident today. The island's very long coastline is shaped by intricate glacier-carved fjords. Flat coastal plains cover a small area. The eastern Canterbury Plains, settled by the English in the 1850s and 1860s, became an important agricultural and sheep-raising region. Today, New Zealand's economy is based largely on agricultural production. It is one of the world's leading exporters of wool and dairy products. This rich land is also endowed with ample resources of timber and natural gas.

Eighteen miles long and more than a half mile wide, the Tasman Glacier, on the eastern side of Mount Cook, is the largest glacier in New Zealand.

The first settlers in New Zealand are thought to have been Polynesians who made long sea journeys from unknown Pacific islands. They apparently called the place Aotearoa, which means "Land of the Long White Cloud," a name that still seems fitting today as we view the land with satellite imagery. In contrast to dry Australia, New Zealand is commonly cloud-covered, and steam and haze often conceal its thermal areas.

The date of the Polynesians' earliest arrival is unclear; theories range from the 10th to the 14th Century AD. Abel Tasman, a Dutch explorer, was the first European to visit New Zealand. In 1642 his landing was thwarted by a violent confrontation with the Polynesians, known as the Maori. He originally named the islands Staten Landt. They were later called Nieuw Zeeland after a province in his Dutch homeland.

The next European to meet Maori warriors was Captain James Cook, the famed British seaman, who circumnavigated the islands and produced the first accurate charts. European settlement began with traders who saw riches to be gained from harvesting whales, seals, and timber. The settlers then introduced guns and new diseases. There were clashes with the Maori. This, along with continued intertribal warfare, began the decline in the Maori population.

Around 1840 British settlement started in earnest. In that year Maori chiefs signed the Treaty of Waitangi, allowing Queen Victoria to annex New Zealand in exchange for the protection of the Crown and for the right of the Maori people to keep their lands and territories. But their lands continued to be taken, and there were frequent periods of warfare. The Maori resistance finally ended by the early 1870s. Decades passed before the Maori population again began to increase. Today the Maori make up around 5 percent of the population of New Zealand.

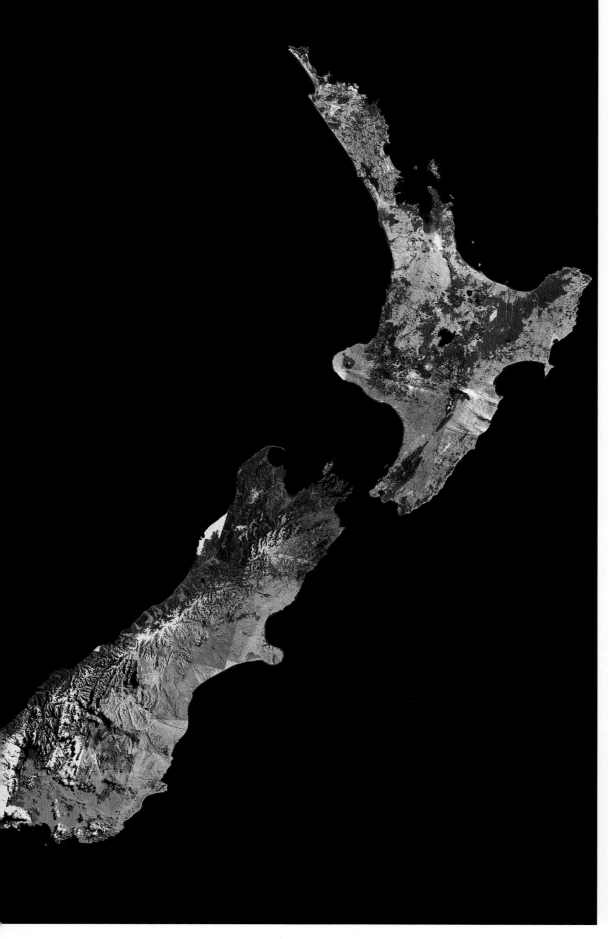

New Zealand

Images for this mosaic were collected over several years (between 1972 and 1980) to produce a clear view of the islands of New Zealand, which are often cloud-covered. Auckland rests on the narrow southern neck of the North Island's northwest peninsula. Two harbors, one deep and one shallow, border Auckland. The peninsula, Northland, is sparsely populated.

The volcanic center of the island is evident around Lake Taupo. Snow-covered Mount Egmont, which last erupted in 1755, rises on the west coast. Mount Egmont and a series of older, eroded volcanoes form the cape that projects into the Tasman Sea.

South Island is mostly mountainous and glaciated. East of the Southern Alps can be seen the grain fields and sheep-grazing pastures of the Canterbury Plains. On the southwest corner of the island is a region called Fjordland, where the dramatically beautiful coast is dissected by deep, glacier-cut valleys, bordered by rugged snow-capped peaks.

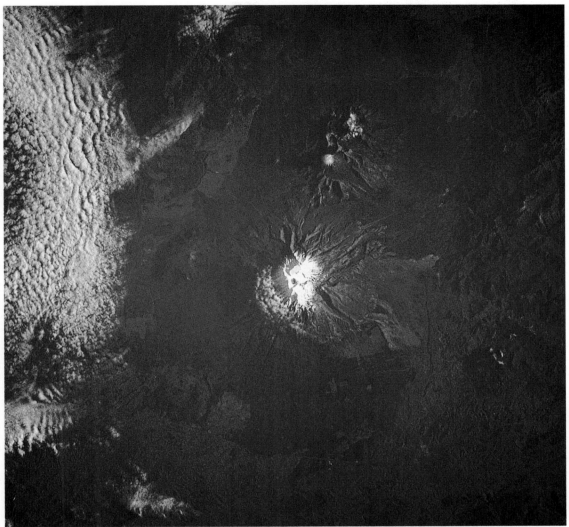

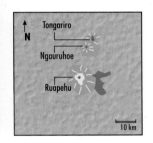

Banks Peninsula

On the east side of South Island is a promontory called the Banks Peninsula (opposite page). Here sturdy volcanic rocks have resisted the erosion that smoothed the rest of the coast. The circular form delineates an old volcano; Akaroa Harbor marks a flooded caldera. Water run-off from frequent rains has formed the radial valleys on the ancient volcano's slopes. The peninsula acts as a barrier to southerly storms in the Canterbury Bight, limiting their spread north.

At the base of the north side of the peninsula lies Christchurch, only partially visible through the clouds. Established in 1850 as a settlement for those of the Anglican faith, Christchurch is the South Island's largest city and a commercial and industrial center for surrounding farmlands.

Live Volcanoes

Three active volcanoes are aligned along a deep fracture running northeast through the center of the North Island (above). The volcanoes lie in Tongariro National Park, built around a gift of land from a Maori chief named Te Heuheu Tukino. In 1886, fearing rival claims and eventual sale of the untouched mountain and forest lands to European settlers and speculators, he presented the land to the Crown as a gift from his people as a way to preserve this natural beauty for the future.

Mount Tongariro, which rises about 1970 m (6,460 ft.), has numerous craters and cones at its summit. On Tongariro's flank rests Mount Ngauruhoe. Relatively young (about 5,000 years old), it has preserved a classic symmetrical cone shape not yet destroyed by erosion. New Zealand's most active volcano, it has erupted more than 60 times in recorded history.

Mount Ruapehu, the North Island's tallest peak, is perpetually snow-covered. But the lake at its summit is heated by the volcano's active interior.

Southern Alps

Abel Tasman, the first European to set eyes on New Zealand, called it "a land uplifted high." The Southern Alps on the South Island are a dramatic part of uplifted New Zealand. The process began during a mountain-building cycle about 2 million years ago in the late Tertiary period. As the Southern Alps were uplifted, the floor of the Mackenzie Basin was depressed, and the basin rapidly filled with debris eroded from the young mountains. The result shows in this image of a craggy land with two-tone lakes.

The three lakes visible in the image occupy troughs that had been excavated by valley-carving glaciers during the Pleistocene period. The lakes' waters are impounded behind glacial moraines, or mounds of broken rock and earth that the advancing glaciers pushed aside and ahead as they moved down the valleys.

Meltwater filled the glacial troughs as the glaciers receded. Eventually the moraines were breached, and outlet rivers flowed out of the lakes. At the outlets of two of the lakes in the basin, Lake Pukaki and Lake Tekapo, dams have been built for the generation of hydroelectric power.

Streams of meltwater flow down from the glaciers, carrying gravel, sand, silt, and clay. The materials, which appear as the dull-shaded sections near the lakes, are deposited in flat or gently sloping areas. These deposits, called outwash trains, fill the valleys above Lake Pukaki and Lake Tekapo.

Remnants of the glaciers that advanced into the basin still exist at the higher elevations in the Southern Alps. The Tasman Glacier lies below Mount Cook (3755 m or 12,316 ft.), the highest point in New Zealand. The mountain, named after the famous explorer Captain James Cook, and the glacier lie in Mount Cook National Park.

The Southern Alps, which extend along nearly the entire length of the island, are so rugged that cross-island railroads and highways are rare. Early travelers compared the island's ranges to Europe's Alps. However, the glacially carved mountains are much more lightly populated.

Islands Born Beneath the Sea

WHERE DO ISLANDS COME FROM? HOW CAN ISLANDS BE SO far away from continents? The answers are complicated and related, in part, to the movement of the Earth's great crustal plates. New material from deep within the Earth is being added onto the plates at mid-ocean ridges or at spreading centers. At those sites, molten rock (magma) works its way to the crustal surface at the ocean bottom. Magma repeatedly fills the cracks or rifts created as the plates slowly move apart.

Islands also form over subduction zones, where one plate is plunging beneath another. This process particularly occurs around the Pacific Rim, where numerous island groups are found. These groups include the Aleutians, the Kuriles, the Ryukyus, the Philippines, the Solomons and New Hebrides, Japan, Taiwan, Tonga, New Guinea, and New Zealand. The long gracefully curving zone that extends southward from Japan to form the Bonin and Mariana island chains, Yap, and the Palau Islands is another example.

Although the subduction process consumes crust, it also generates volcanoes and deposits that build landmasses, as in the southern Mariana Islands. Subduction can also build landmasses such as Japan. Such "island arcs," viewed from space, are among Earth's more magnificent features. Commonly, arc volcanoes become the foundations of more complex islands where reef growth and sedimentation combine to produce increasing mass and diverse rock types. Guam and other southern Mariana Islands are examples of this.

Other islands, such as those of the Hawaiian chain, occur in the deep ocean with no apparent relation to spreading centers or subduction zones. These islands are probably formed where hot magma has squirted upward, in the direction of least resistance, from deep in the Earth's mantle, burning through the crust to form great volcanic piles of lava and ash. Such activity occurs in pulses at intervals of hundreds of thousands to millions of years. But the crustal plates are mov-

ing, so that each major pulse builds a volcanic pile in a new place, directly over the "hot spot." A chain forms in the direction of plate movement. This phenomenon is well illustrated by the string of islands from Hawaii itself west-northwestward to Midway, about 2600 km (1,600 mi.) away.

Beneath the heavy volcanic piles, the oceanic crust gradually yields, and the volcanic masses subside. Those masses, already well below the sea, go deeper. Masses at or above the surface generally are eroded before disappearing below the water. Some of them, pared down by waves, form flat-topped seamounts, or guyots, in the depths of the ocean.

Another kind of volcanic mass stays above or just below the surface long enough for a fringe or crown of reefs, composed primarily of calcium carbonate and comprising the skeletons of animals and plants, to form. Their remains merge to become wave-resistant structures. Such reefs can form only in warm, relatively shallow water where some type of base or platform is available upon which to grow.

Typically, reefs begin as fringes around the shoreline of a volcanic island. The reefs gradually separate from the sinking island mass and evolve toward a barrier reef, encircling both a lagoon and what remains of the volcanic pinnacles. In some places such reefs may continue growing upward for thousands of feet as the volcanic bases subside. These eventually form atolls—sinuously distorted, or finely articulated, rings of reefs commonly displaying small, low islands. Well-developed atolls characterize the Marshall, Gilbert, and part of the Caroline islands.

Today Pacific islands can be seen in all stages of development, from reef-fringed volcanoes to subsiding volcanic masses ringed by lagoons and barrier reefs to atolls rising only a few feet above sea level in most cases. These ultimate reef complexes mark the places where great volcanic edifices were formed. (D.B.D.)

Charles Darwin selected Bora-Bora 150 years ago as his classic example of a subsiding "barrier reef island." Bora-Bora has played a key role in scientific argument concerning the development of the coral reefs of the world, and Darwin's view still persists.

Pagan Island

The largest of the northern Mariana Islands, Pagan Island is actually a bundle of small volcanoes. The largest is shown in eruption (left), with its plume of ash and dust. In 1980 the forerunner of the eruption covered the middle of an airstrip with about 4 m (13 ft.) of basaltic lava. Pagan is about 520 km (322 mi.) north of Guam and 2390 km (1,482 mi.) south of Tokyo.

Pagan and its neighbor to the south, Alamagan, are generating cloud trains. These result from the condensation of water vapor, owing to cooling where a humid air mass flows upward and over the summit of each island.

These islands are part of the Mariana Arc, the northern reach of which includes young volcanoes, thought to be less than 2 million years old. During World War II Pagan was a Japanese air base with 8,000 personnel. Today Pagan supports about 35 farmers and fishermen. (D.B.D.)

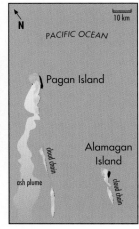

Raiatea, Tahaa, Bora-Bora, Tupai

This group of islands (below) in French Polynesia displays a striking sequence of sinking volcanic masses, formed by deep-ocean "hot-spot" activity. They illustrate a progression of island development, beginning with great piles of volcanic flows and ash and ending with coral reefs that mark what was once a high island. Northernmost is Tupai, the small atoll. Bora-Bora sits near the middle of the picture. About 24 km (15 mi.) southeast of Bora-Bora are Tahaa (seen in its entirety) and Raiatea (partly out of the picture). They both lie within a single enclosing reef.

Known to the western world since Captain Cook's visit of 1769, Bora-Bora, Tahaa, and Raiatea are the high islands of the Leeward group of the Society Islands. All have steep cliffs of volcanic rocks—chiefly lava—and rugged peaks.

Narrow marine benches provide space for houses and gardens in some places. But the populations have not increased in proportion to that of Tahiti, the cultural hub and headquarters of the French administration in the nearby Windward Islands 240 km (150 mi.) to the east-southeast. The people of these Leeward Islands have retained an independent attitude, based partly on a history of rebellion in opposition to French rule of the Societies.

Geologically, the picture suggests an ideal progression from the highest islands to an atoll. Raiatea and Tahaa have relatively narrow reefs that have not changed completely from fringing structures to barriers. Tahaa has subsided slightly more than Raiatea and thus has broader lagoons separating the reefs from the volcanic shorelines.

Bora-Bora is at least tens of thousands, and possibly hundreds of thousands, of years further along in the progress of subsidence and the biological mechanisms of reef development. The island displays a well-developed barrier reef on all sides and broad reef flats. The atoll, Tupai, marks the final vestiges of an old volcanic island that has subsided deep enough to become covered entirely by reefs and off-reef or lagoonal sediments.

Passages through the reefs surrounding the high islands are maintained in most places by the outward flow of terrestrial streams, whose fresh water is lethal to the corals and other reef organisms. Low islands surmounting the reefs are considered remnants of younger reef growth up to an old sea level. (D.B.D.)

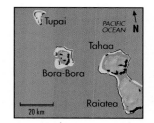

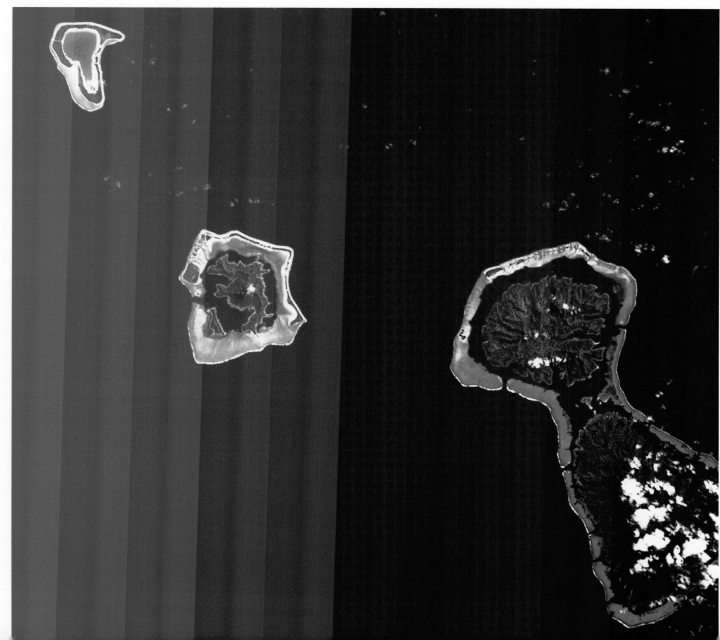

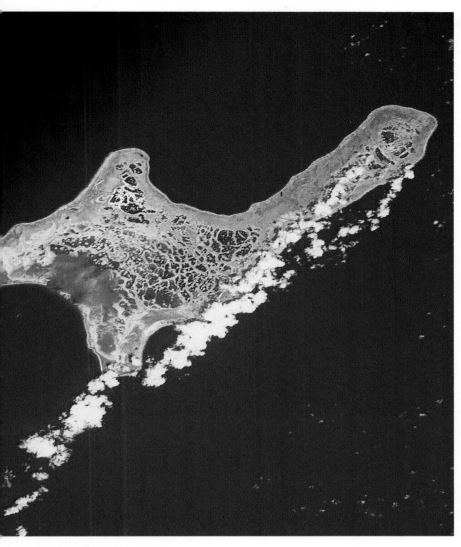

(following pages)

New Caledonia

Mountainous New Caledonia is one of the largest islands in Oceania. Unlike many of the Pacific islands that have volcanic origins, it is composed of metamorphic and sedimentary rocks.

The southeast trade winds bring rain to the eastern shore and slopes of the island. The western side is relatively drier, due to the "rain shadow" effect created by the mountains. Vegetation reflects the uneven rainfall: conifers on the eastern side, savanna grasslands on the western side.

In this SPOT scene, the eastern mountain slopes are bright red, indicating healthy vegetation. The floodplains and alluvial deposits along the rivers on the west coast have been cultivated, and these also appear bright red.

New Caledonia is surrounded by barrier and fringing reefs. These are breached by channels, often near river mouths. The fresh water discharge limits coral growth along these channels. Mangrove swamps on the west coast appear as patches of dark red and gray.

Kiritimati

The largest atoll in the Pacific, Kiritimati is in the Line Islands about 2000 km (1,200 mi.) south of Hawaii. Kiritimati was formerly called Christmas Island, the name given it by Captain James Cook, the British navigator and explorer. He first encountered the island on Christmas Eve 1777, during his last voyage. Cook and his crew remained on the island for about ten days. Serving as an officer with Cook on that voyage was William Bligh, who later gained notoriety as captain of the Bounty during its mutiny.

In the first half of the 19th Century the island was home to whalers. It later developed a thriving trade in copra (dried coconut meat). After World War II, Christmas Island was the site of nuclear-weapon testing by both Great Britain and the United States.

Kiritimati, a fish and bird sanctuary, also supports a small tourist industry. Once part of a British colony, Kiritimati became a member of the Republic of Kiribati in 1979. Kiribati, with a population of 64,000 (1985), is composed of numerous islands including the Gilberts, most of the Line Islands, and the Phoenix Islands of the central Pacific.

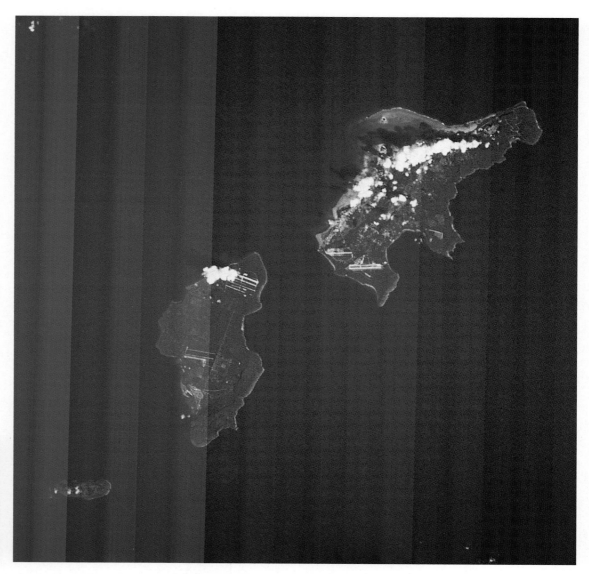

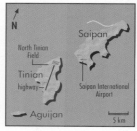

Tinian and Saipan

Built on volcanic foundations perhaps 50 million years old, Tinian (center) and Saipan (upper right) islands have developed by the growth of coral reefs and associated sediments on centers not very different from present-day Pagan, about 330 km (205 mi.) to the north. Through millions of years of successive submergences and emergences, the island masses have increased to their present sizes. In the rich soil of their broad, gentle slopes grow sugar cane, bananas, pineapples, lemons, limes, oranges, coconuts, and other foods. Southwest of Tinian is Aguijan, an island whose stepped terraces represent former sea levels.

A prominent highway can be seen along the length of Tinian. Clouds are visible forming along the peaks of Saipan and over the broad low terrace of north Tinian.

Archaeologists have turned up artifacts including potsherds, axes, and chisels indicating that these islands were inhabited at least 3,000 or 4,000 years ago. In more recent history, Saipan was a port of call for Spanish galleons en route from Manila to Mexico. Treasures have been retrieved from a galleon sunk just off of Saipan. Possession of the islands passed from Spain to Germany in 1898, and then to Japan after World War I. The Japanese colonized intensively and established large-scale sugar cane farming as well as the mining of manganese and copper on Saipan and phosphate on Tinian.

In 1944 U.S. Navy and Marine forces captured Saipan and Tinian after intense fighting. On Saipan, all but about 1,000 of the 30,000 Japanese defenders died. Some killed themselves, as did many civilians who threw themselves off cliffs as the battle for the island ended.

Airfields for B-29 Superfortress bombers were built on both captured islands. North Tinian Field, with six runways nearly 4.8 km (2 mi.) long, became the largest airport in the world. The northernmost of North Tinian's runways, still in use during military exercises and obvious on this photograph, is the one from which the B-29 named "Enola Gay" took off to carry the atomic bomb dropped on Hiroshima. Saipan and Tinian each, briefly, had a population of 250,000 U.S. military personnel during the buildup for assaults on Iwo Jima and Okinawa.

Today Saipan International Airport, a former B-29 base, accommodates many Japanese tourists who come to see the old battlefields and war memorials. Tinian has been used extensively for cattle ranching and will be the site of new casinos and hotels. Aguijan is being resettled by people from Tinian. (D.B.D.)

Ulithi Atoll

Oddly shaped Ulithi lies at the western end of the Caroline Islands, which are located about 480 km (300 mi.) southwest of Guam, at the south end of the Mariana chain. The Carolines include high islands, such as Truk, Ponape, and Kosrae, as well as a number of atolls like Ulithi.

Unlike Bora-Bora or Raiatea, Ulithi has an enigmatic form characterized by the tortuous configuration of its reef. Its sinuosity is interrupted by sharp angles. Many geologists believe that uneven subsidence, tilting, and faulting of the volcanic platforms affected the ultimate shape of the atoll reefs and the enclosed lagoonal areas.

The prominent island outside the closed figure of the reef is Falalop, which supports a small population. Immediately west-northwest is Asor. South-southeast of Falalop are Pau and Losiep astride their own reef, detached from the atoll itself.

U.S. military forces captured Ulithi in 1944. Although strategists saw it as an advanced fleet base, it was the northernmost island of the Ulithi Atoll, Mogmog, that thousands of U.S. Navy sailors remember best. Mogmog was the first dry land they set foot on after long months at sea. Here they found sports arenas, playing fields, movies, a base exchange, post office —and cold beer.

Ulithi had a more serious role, though, in the spring and summer of 1945. Here was a final staging area for the invasion of Japan, scheduled for November 1, 1945. Great concentrations of ships anchored in the lagoon, including probably the largest number of U.S. Navy aircraft carriers ever converged in one place during tactical operations. The final task forces were to assemble at Ulithi prior to steaming north to Kyushu, the southernmost of the four Japanese home islands, for the planned amphibious assault. The abrupt end of the war in August 1945 found many ships and aircraft still on their way to Ulithi. Most Navy personnel, who had never heard of the atoll, thought the Micronesian name was some sort of code word.

In 1952, geologists S. O. Schlanger and J. W. Brookhart of the U.S. Geological Survey concluded that comparisons of Asor and Falalop provided clear evidence of the height of former sea levels. Ocean waters stood at 2 m (7 ft.) and later at less than a meter (2 to 3 ft.) above today's sea level, probably not more than 3,000 years ago. (D.B.D.)

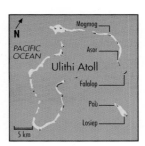

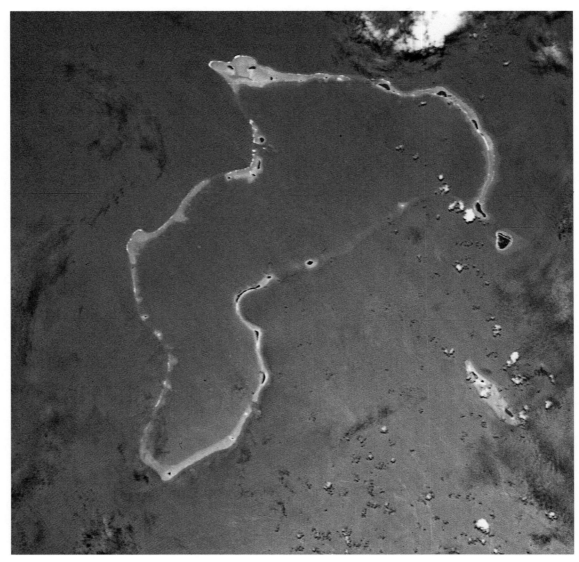

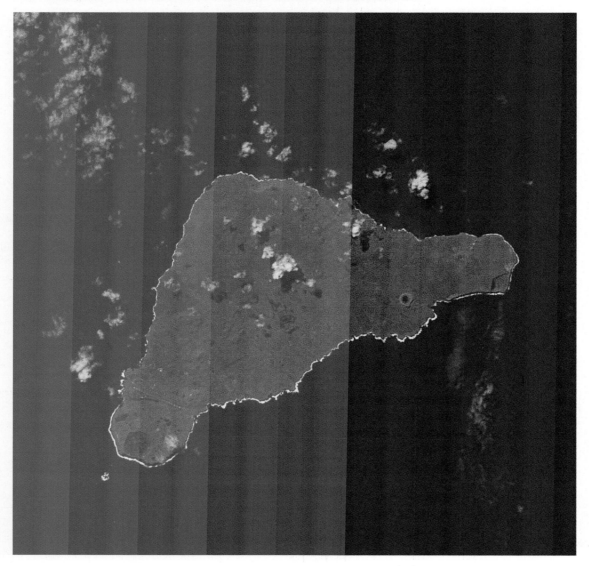

Easter Island

Isolated Easter Island is about 2000 km (1,240 mi.) from the nearest inhabited island, Pitcairn, and more than 4000 km (2,480 mi.) from the coast of South America. Easter Island

was formed by several now-dormant volcanoes. Terevaka, at 600 m (1,968 ft.) the highest volcano on the island, is partially obscured by clouds in this Space Shuttle photograph (below left). An airport runway, northeast of the Rano Kao volcano, has been extended for use as an emergency landing strip for the Space Shuttle.

Some 3,000 people were living on the island when Europeans arrived on Easter Day in 1722. By 1877, all but 115 islanders had died of European disease or had been carried away as slaves. Since 1888 the island has been a province of Chile.

Long before the Europeans' arrival, islanders quarried dark volcanic rock and carved it into about 600 large statues (above, left), ranging from 6 to 9 m (20 to 30 ft.) high. Many are mounted on basalt platforms. The quality of the stonework is equal to that found in Inca ruins of Peru. This has helped to inspire theories of westward migration from South America into Polynesia.

Polynesians probably reached Easter Island from the west around 400 AD while they were on a voyage of exploration. The Polynesians paddled large, seagoing canoes and navigated by the stars, and current, wind, and wave direction to determine position, speed, and direction of travel. They settled Pacific islands from New Zealand to Hawaii.

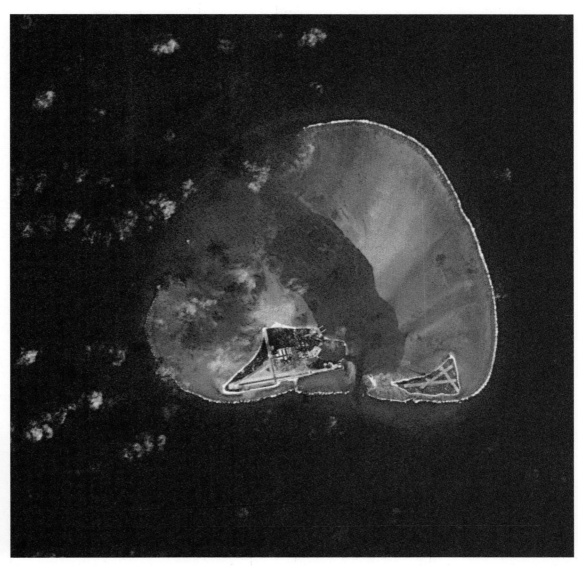

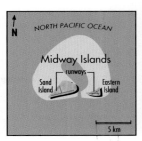

NORTH PACIFIC OCEAN

Midway Islands

Sand Island — runways — Eastern Island

5 km

Midway Atoll

An emerald dot in the middle of the northern Pacific Ocean, Midway Atoll consists of two low islands and a surrounding reef. These are remnants of an extinct, subsided volcano that is part of the chain of volcanic islands, which includes the Hawaiian Islands. The atoll lies near the international dateline.

Midway, uninhabited when it was discovered in 1859 by an American ship, was an-

nexed by the United States in 1867. Sand Island became an important transpacific communications and transportation link, beginning with the construction of a cable station in 1905. A commercial seaplane base was built at Midway in 1935 for transpacific flights by Pan American Airways Clippers. In 1941 the U.S. Navy built a harbor at Sand Island and an airfield on Eastern Island. Midway's location made it strategically

important for the defense of Hawaii against the Japanese.

In June 1942 a major battle here ended in a decisive victory for the United States. U.S. aircraft destroyed four Japanese aircraft carriers, sank a cruiser, and badly damaged another cruiser and two destroyers. Japan has never mounted another major offensive campaign in the Pacific.

The airfield on Eastern Island is no longer in use and has been replaced by two

longer runways on Sand Island. During World War II 2,000 U.S. sailors and marines were stationed at Midway. Fifty military personnel now operate the Naval Air Facility at Sand Island.

Midway Atoll has a large population of gooney birds, a species of albatross. The atoll is maintained as a National Wildlife Refuge under an agreement between the U.S. Navy and the U.S. Fish and Wildlife Service.

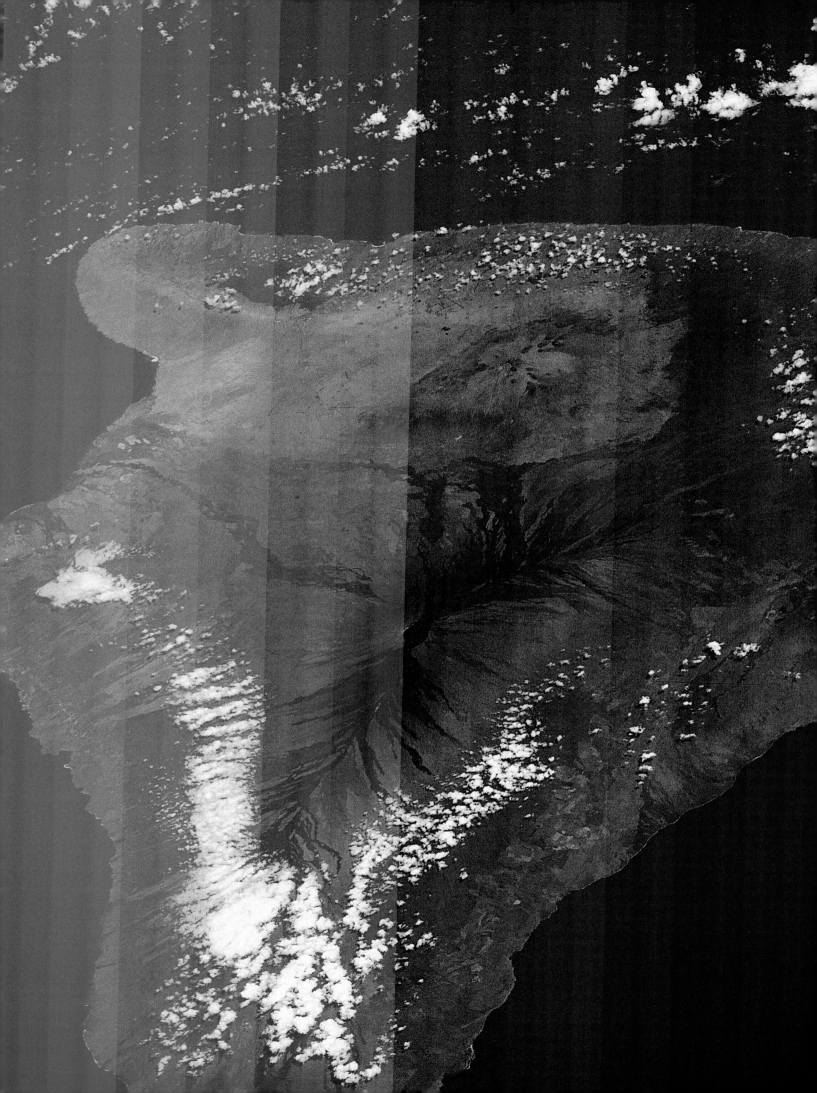

Hawaii

The Big Island, Hawaii (opposite page), is the largest and youngest of eight major islands that make up the 50th state of the United States. A traveler standing on the steep cliffs of the island's South Point, or Ka Lae, has reached the southernmost point in the United States.

The island is the top of a huge mountain made up of five volcanoes whose summits rise above sea level. Two, Kilauea (above) and Mauna Loa, are active and comprise the Hawaii Volcanoes National Park.

The island's volcanic origins are clearly evident in the image to the left. Dark lava flows snake down the flanks of Mauna Loa and Kilauea. Both are capped by large collapsed calderas (craters). To the north lies Mauna Kea, inactive for thousands of years. Hawaii's highest peak, it rises to 4205 m (13,796 ft.). Because of its great height and clear air it has become the site of several astronomical observatories.

The towering summit of Mauna Kea is often snow-covered. Despite the thin air, skiing is popular. Elsewhere on the island, in contrast, are lush tropical vegetation, warm black-sand beaches, fields of sugar cane, and a thriving orchid industry.

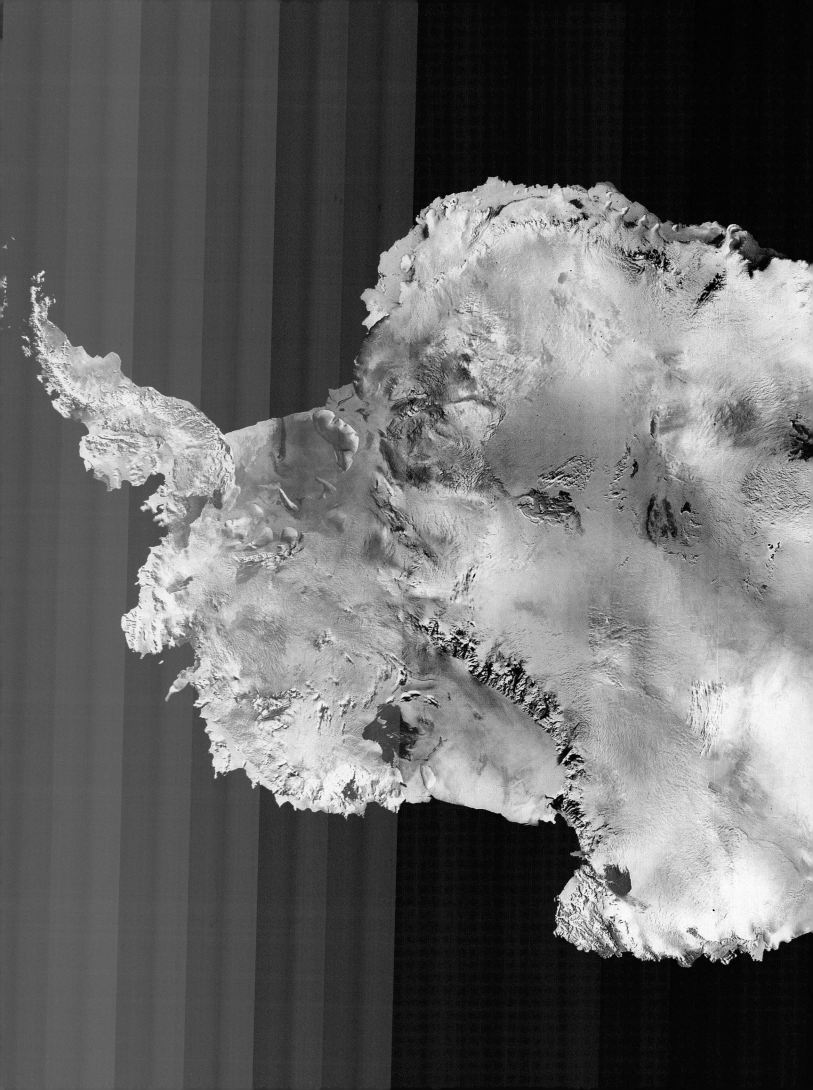

A Continent of Ice

TWENTY-THREE SATELLITE SCENES, COLLECTED BETWEEN 1980 AND 1983, produced this mosaic of Antarctica. The prominent Antarctic Peninsula juts northward to within 1000 km (620 mi.) of Cape Horn, the southern tip of South America. The peninsula was formed during the same orogeny, or mountain-building process, that raised the Andes. The two nearly straight sections of the Antarctic's perimeter are the ice fronts of the Ronne Ice Shelf (upper left) and the Ross Ice Shelf (lower center). Ice fronts, the steep seaward margins of the floating ice shelves, range in height from 2 m to more than 50 m (6 to 160 ft.).

The rugged Transantarctic Mountains (lower center) are crossed by a number of large outlet glaciers that drain ice from the East Antarctic Ice Sheet into the Ross Ice Shelf. These outlet glaciers provided routes for explorers Roald Amundsen, Robert Falcon Scott, and Sir Ernest Shackleton (see page 290) to ascend from the Ross Ice Shelf to the Polar Plateau.

The collection of satellite imagery over Antarctica is not an easy task. Many spacecraft orbits do not reach high enough latitudes to get images of the polar regions. Satellites that do cover the poles often carry only low-resolution instruments. Heavy cloud cover regularly obscures scenes, and the cloud cover is difficult to distinguish from snow. The high reflectivity of snow and ice can saturate sensors; the result is images that are mostly featureless and white. Computer processing can often improve the contrast and quality of the imagery, but only if significant information is contained in the raw data. When high-resolution satellite imagery of Antarctica is possible, scientists get the means to study areas that are inaccessible and hard to traverse. Broad geologic structures can be defined and measured. Changes in the landscape such as those caused by glacial flow or retreating ice shelves can be documented by comparing images collected at different times. Areas that took early explorers weeks to cross now can be surveyed and mapped with only a few images, recorded in but a few minutes.

About 90 percent of the Earth's fresh water is locked up as ice in Antarctica's glaciers, ice sheets, and ice shelves. Eighty to 95 percent of the solar energy arriving at the surface of Antarctica is reflected back into space by the continent's snow and ice. Because of the planet's spherical shape, the polar regions receive less of the sun's energy per unit of surface area than regions in lower latitudes. The angle at which solar energy reaches the Earth's surface becomes more oblique as latitude increases. The decreasing angle of the sun's rays and the high reflecting property of ice are major factors in making polar regions cold in comparison to temperate and tropical areas at lower latitudes.

Cold, dry air descends over the Antarctic, creating a system of high atmos-

AVHRR MOSAIC, IMAGE SUPPLIED BY
NRSC LTD., WITH THE AGREEMENT OF
NRSC, FARNBOROUGH, UNITED KINGDOM

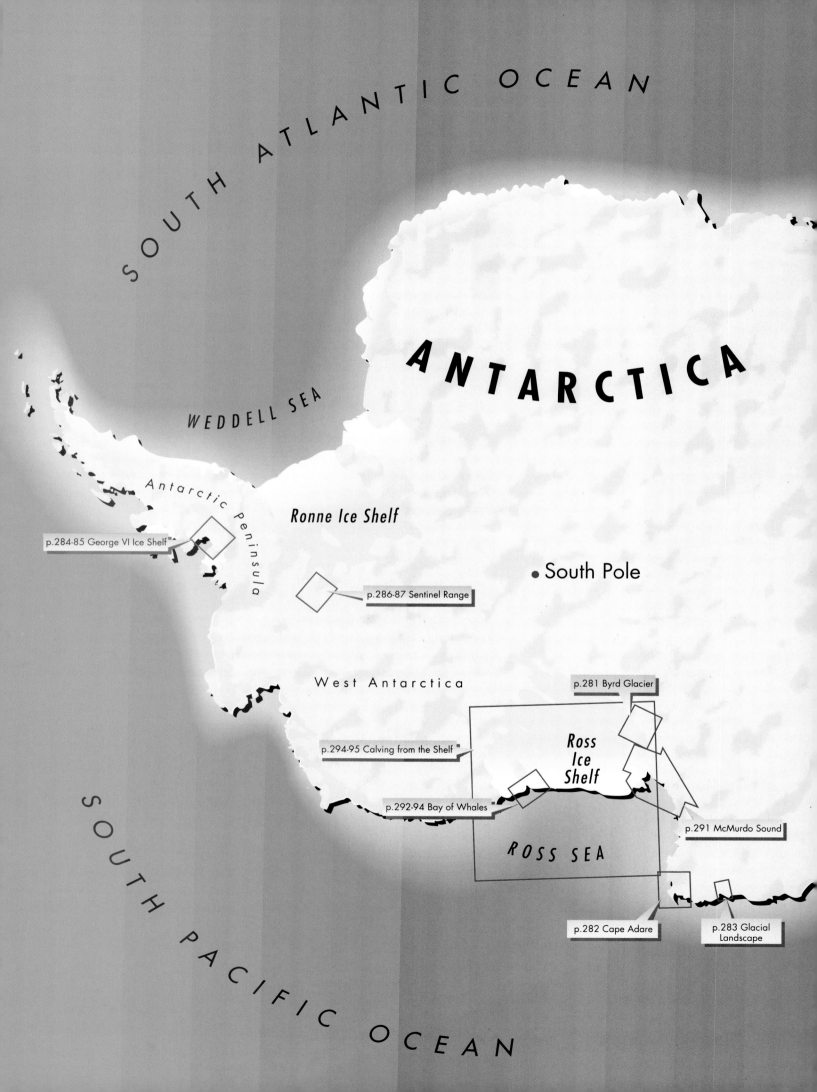

SOUTH ATLANTIC OCEAN

SOUTH PACIFIC OCEAN

WEDDELL SEA

ANTARCTICA

Antarctic Peninsula

Ronne Ice Shelf

South Pole

West Antarctica

Ross Ice Shelf

ROSS SEA

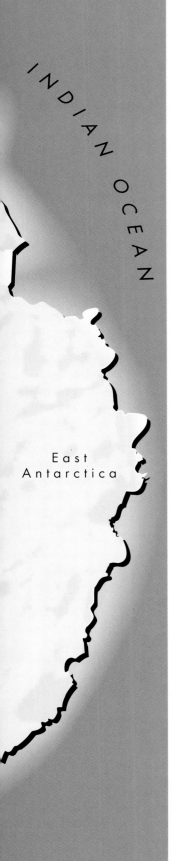

INDIAN OCEAN

East
Antarctica

pheric pressure that dominates the continent's interior climate. The coldest temperature on record at the Earth's surface, -88.3°C (-126.9°F), was measured at the Vostok Research Station, about 1300 km (800 mi.) from the South Pole.

The combination of extremely low temperatures and high pressure produces an atmosphere that contains about one-tenth of the water vapor usually found at mid-latitudes. Precipitation at the South Pole averages less than 51 mm (2 in.) per year, qualifying the Polar Plateau as one of the Earth's great deserts.

Antarctic fauna concentrate along the coastal margins of the continent where the climate is relatively moderate. The only land animals are small invertebrates, such as mites and thread-wide worms called nematodes. Offshore waters abound with marine life, particularly krill, a small, shrimp-like crustacean that humpback whales feed on, as do minke whales, fin whales, and blue whales. Krill is also consumed by fish, squid, seals, penguins, and seabirds.

Twelve species of whales visit Antarctica's waters, and eight species of seals also find their way to Antarctica's coasts. Forty-three species of birds are known to breed on the continent, including albatrosses, petrels, gulls, terns, cormorants, ducks, and five varieties of penguins.

In 1956 a U.S. scientific team landed at the South Pole to spend the winter and conduct experiments for the International Geophysical Year (I.G.Y.) Since then the U.S. polar station has been continuously occupied. From 1957 to 1958 scientists from 12 nations launched expeditions in Antarctica, setting up more than 50 stations as part of the I.G.Y. The studies focused on meteorology, oceanography, upper-atmosphere phenomena, seismology, and glaciology. At that time, about half of Antarctica had never been seen.

Cooperation among scientists from many nations led to an international agreement guaranteeing free access to the continent. The 30-year, 1961 Antarctic Treaty specified that the continent was to be used only for peaceful purposes and that scientific information obtained there was to be exchanged freely. In 1991, the Madrid Protocol (to the 1961 treaty), agreed to by 39 countries, established a 50-year moratorium on the exploration and exploitation of Antarctica's natural resources. However, the treaty is only binding on the signatory nations, and more than 100 countries have not signed. A number of countries, including some that signed the treaty, maintain territorial claims in Antarctica.

An example of the value of scientific studies here came in 1969 when a Japanese team discovered nine meteorites on a small patch of ice. The meteorites' characteristics suggest that they originated from a large planetary body with relatively recent geologic activity. They have been classified with a group of meteorites that some scientists believe arrived on Earth from Mars. The gas concentrations trapped in one of the samples match those measured in the Martian atmosphere by NASA's Viking mission to Mars.

Some scientists theorize that if a large meteor hit Mars, rock fragments could be ejected with enough force to escape from the planet's gravity and eventually find their way to Earth. Such meteorites hold the promise of answering questions about the nature and history of Mars, posing an intriguing irony: in one of the last places to explore on Earth may be found some of the first pieces of Mars—the next goal, perhaps, in the story of human exploration.

ANTARCTIC ICE

ALTHOUGH MOST AREAS OF THE EARTH HAVE BEEN VISITED and charted, the icy continent of Antarctica remains largely unexplored and poorly mapped, a unique and remote region with unfamiliar landscapes and shifting outlines. The subglacial terrain is covered with more than 11.7 million sq. km (4.5 million sq. mi.) of thick ice sheets. In some places, the sheets are more than 4 km (2.5 mi.) deep.

The ice sheets conceal the underlying continental bedrock, only about 2 percent of which is visible. Most of the rocky surface is seen only as small hills, called nunataks, that protrude through the ice. Although the ice sheets mask the true structure of the continent below, their contours mimic the broad topography of the subglacial terrain. And they exhibit their own distinctive features, such as ice rises, crevasses, and other glacial landforms (see pages 281, 283, and 286-87).

Glaciers form from the thick accumulation of snow that compacts and recrystallizes as ice. Although it is natural to think of ice as brittle and unyielding, under the pressure of its own heavy layers, ice will flow. Generally, glacier movement is slow (only a few centimeters a day), but rates vary: observers have recorded rapid surges. Mountain glaciers originate in highland snow fields and follow stream valleys. Continental glaciers are broad ice sheets that move outward in all directions.

Glaciers shape the terrain through erosion, deposition, and the transportation of large amounts of material. As glaciers advance, they scour land surfaces and valley walls and carry embedded rock fragments along with them. Glacial meltwaters deposit clay, sand, and rocky debris as they flow out from the glacier.

In addition to the ice blanketing most of the land surface, Antarctica possesses vast fringing ice shelves. These are anchored to the coast but float on the ocean waters. (See pages 284 285, 292-93, and 295.) Ten percent of Antarctic ice is found in small shelves all around the continent and in

Basic winter mountaineering skills are taught to researchers who work at many of the stations in Antarctica. Scott Base is located on Ross Island, which is crowned by the volcano Mount Erebus. Clouds shroud the summit of Erebus (above), about 40 km (25 mi.) from the climber ascending a short icy slope. Crampons, metal teeth attached to the bottom of mountain boots, and ice axes make ice climbing possible.

three immense shelves that hug the coast in embayments of the Ross and Weddell seas. These shelves can be several hundred meters thick and are fed by glaciers flowing out toward the edge of the Antarctic landmass.

The Ross Ice Shelf, the largest in the world, alone covers an area larger than the state of California. Discovered in 1841 by James Clark Ross, a British polar explorer, it provided a flat and accessible, though often treacherous, path to the Antarctic interior for several explorers.

Large amounts of sea ice are also found in the ocean waters around Antarctica. This ice takes the form of icebergs, which calve into the sea from the edges of glaciers and ice shelves, and pack ice, which forms from the freezing of sea water. The flat-topped Antarctic tabular icebergs can be huge, commonly reaching several kilometers across. Icebergs more than 100 km (62 mi.) long have been sighted. Pack ice forms a band around Antarctica that moves generally east to west with prevailing currents. Its area fluctuates with the changing seasons (See pages 282, 289, and 293.)

The ice of the Antarctic has global significance. Serious debate about Earth's warming has raised questions about the possibility of the shrinking of the Antarctic ice sheets. If sufficient amounts of Antarctic ice were to melt, sea level would rise, flooding low-lying coastal areas across the world. Global climate patterns could also be affected.

Concern about climatic change has been raised particularly by scientists' continuing observations of two phenomena. For some while, there has been shrinking of ice shelves along the curving Antarctic Peninsula. There have also been large calvings from the edges of the Ross and Filchner ice shelves. While the calvings appear to be cyclical occurrences that happen naturally as ice shelves grow, further study is needed to determine if the other losses are natural or caused by human activities.

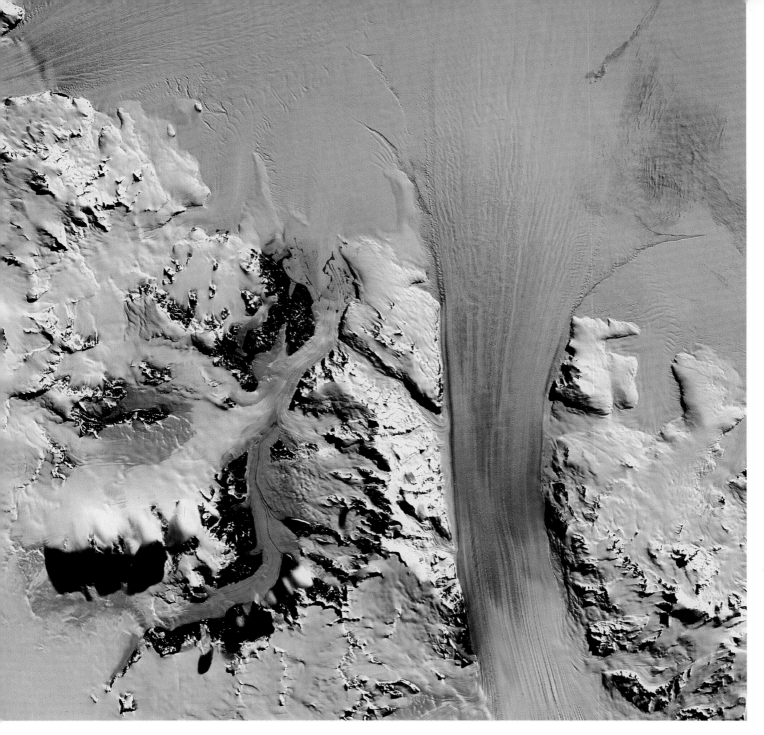

Byrd Glacier

Many glaciers flow through the valleys of the rugged Transantarctic Mountains onto the western edge of the Ross Ice Shelf. The largest of these, Byrd Glacier, is more than 20 km (12 mi.) wide. Outlet glaciers like Byrd flow under their own weight through pre-existing subglacial valleys, scraping, eroding, and shaping the terrain as they go.

Byrd Glacier travels more than 100 km (60 mi.) through the mountains. Long ridges and furrows are visible parallel to the valley walls. Crevasses 20 m (about 70 ft.) deep are also found in the valley. Byrd is one of the fastest moving glaciers of its kind in Antarctica, advancing several hundred meters per year. Where it

flows into the Ross Ice Shelf, crevasses and sharply angled rifts reflect the stresses caused by the interaction of the two bodies of ice. Here, the floating shelf is nearly broken apart by the force of the glacier advancing across it. To the northwest, the narrow Darwin and Hatherton glaciers discharge much more slowly into the Ross Ice Shelf.

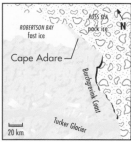

Cape Adare

Two types of sea ice are separated by Cape Adare. Fast ice is attached to the coast, whereas pack ice is drifting sea ice. Fast ice fills Robertson Bay, but pack ice drifts in the north Ross Sea (top center). In 1895 two Norwegians, Carsten Borchgrevink and Leonard Kristensen, became the first men to land on the Antarctic mainland by coming ashore at Cape Adare. Unable to journey inland, they sailed south to the Ross Ice Shelf. The explorers found that the edge of the ice shelf had moved a substantial distance south since its discovery in 1841. Borchgrevink set out for the South Pole on skis at the Bay of Whales (see pages 292-93) and created a new record for an approach.

More than a decade of other failed attempts passed until Roald Amundsen succeeded in reaching the goal on Dec. 14, 1911. (Borchgrevink returned to Cape Adare in 1899 with a nine-man British-Norwegian team. This expedition was the first to winter over on the Antarctic mainland.)

The Tucker Glacier (bottom left to right) crosses the Borchgrevink Coast to discharge into the Ross Sea. The tributary glaciers flowing into the Tucker Glacier follow right-angle faults and joints in the underlying rock, creating a rectangular drainage pattern. This 170-km (100-mi.) glacier system, though quite extensive, is supplied by local accumulation of snow and does not drain the East Antarctic Ice Sheet.

Glacial Landscape

Complex flow lines and crevasse patterns can be seen on the surfaces of the glaciers in this image collected along the Pennell Coast by the Landsat sensor known as the RBV (Return Beam Vidicon). The general usefulness of the RBV sensor is limited by the small number of distinctive images available. But the sys-tem carried on the Landsat 3 satellite has a high resolution (30 m or 980 ft.) that captures the fine detail of the Antarctic ice surfaces and allows their intricacies to be sharply delin-eated. (For more about satellite systems, see pages 298-301). Long shadows also accentuate some features and are due to the low solar elevation angle (6° above the horizon) at the time the image was recorded.

Along the length of the glaciers, flow lines run parallel to the direction of the ice's ad-vance. A complex pattern of rifts and crevasses has formed on the tongue of the Lillie Glacier where it coalesces with the George Glacier before flowing across the edge of the continental landmass and into the ice off the coast (fast ice).

Crevasses also occur where the Lillie and Graveson glaciers meet. The structure marking the confluence of these glaciers may reflect a topographic fea-ture in the bedrock hidden by the ice. The Anare Mountains and Bowers Mountains rise above the icy landscape. Small valley glaciers, originating in the Bowers Mountains, feed the Graveson Glacier below.

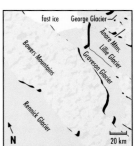

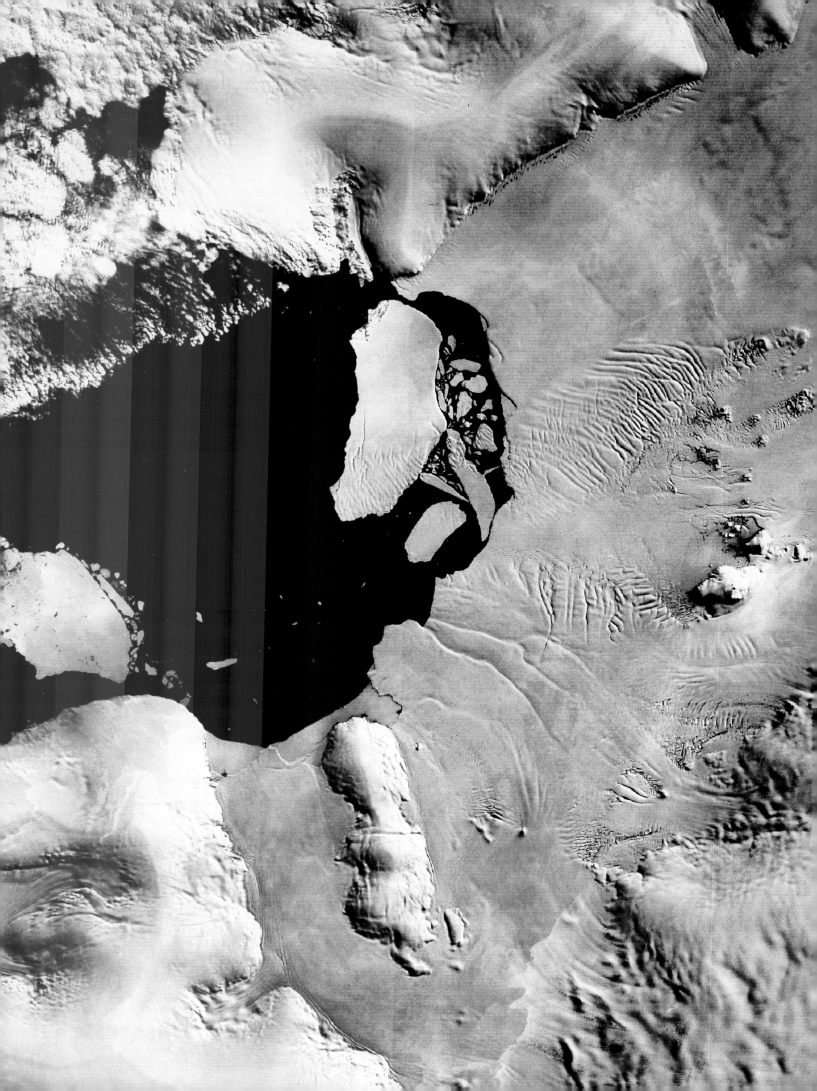

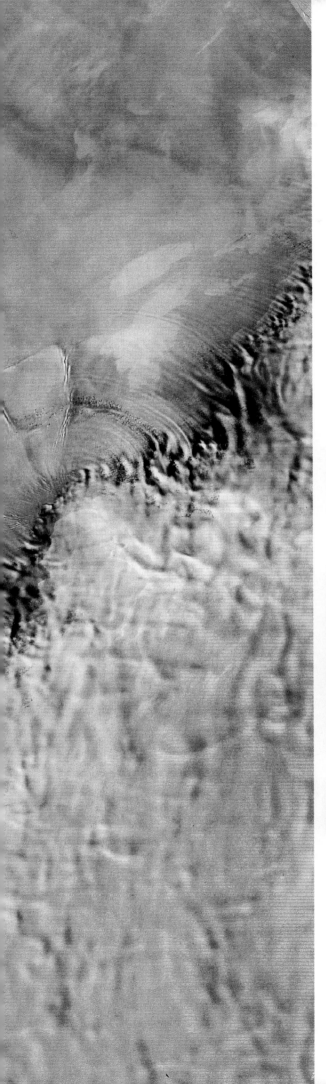

George VI Ice Shelf

In the Antarctic most icebergs are formed when ice calves, or breaks off, from ice shelves. Antarctic icebergs (below) break off from a shelf and produce broad tabular bergs, unlike Arctic icebergs that calve off from glaciers and have irregular shapes. A tabular iceberg has recently calved off the George IV Ice Shelf and is adrift in the Ronne Entrance in this January 1973 Landsat scene (left).

The Ronne Entrance is a polynya, an area of open water surrounded by ice. Local winds keep the Ronne Entrance ice-free for most of the year. Ice floes composed of drifting sea ice are also floating in the Ronne Entrance. During the austral summer a string of meltwater ponds (top center) forms on the boundary of the ice shelf and Alexander Island.

Crevasses (upper center) are visible on the ice shelf and are formed downstream of a point where the shelf is grounded on a shoal. Spaatz and DeAtley islands (lower left) are ice rises located off the English Coast on the western side of the Antarctic Peninsula (lower right). Ice rises, which are small ice caps covering islands or larger shoals, occur within or along the boundaries of ice shelves. In this scene it is easy to distinguish the Antarctic mainland (lower right) from the adjacent ice shelf. But it is difficult to discern the precise coastline beneath the layers of snow and ice.

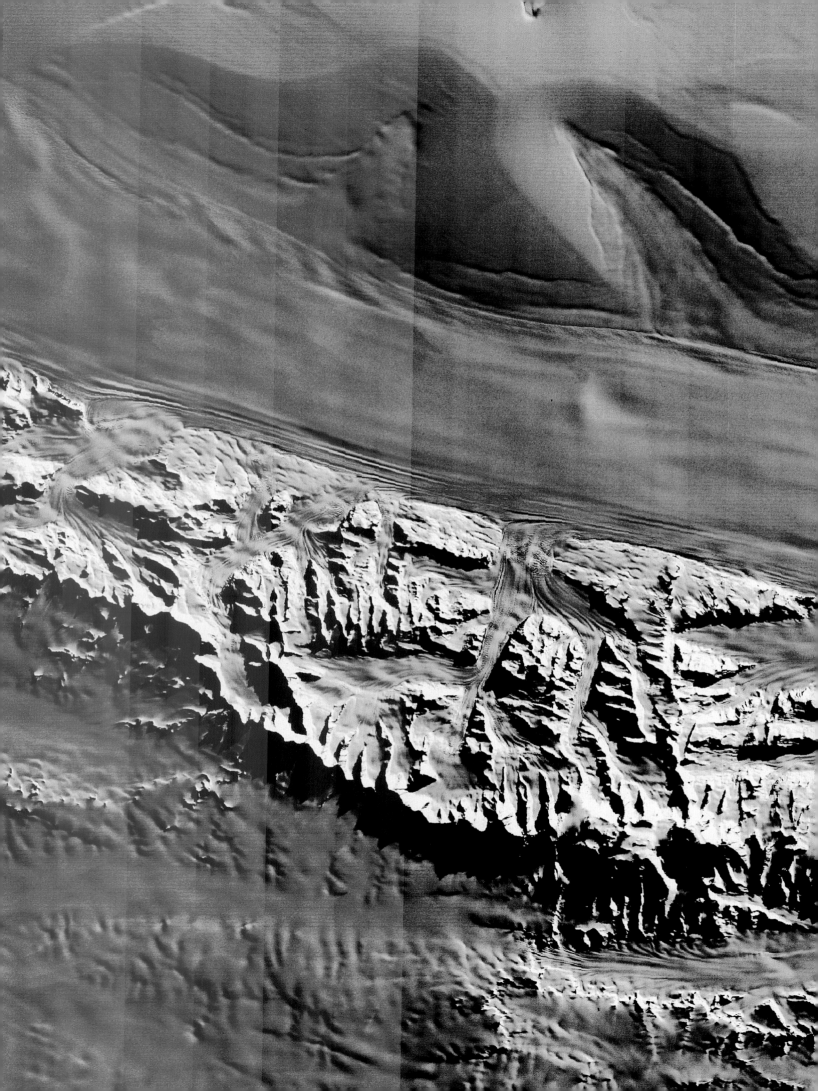

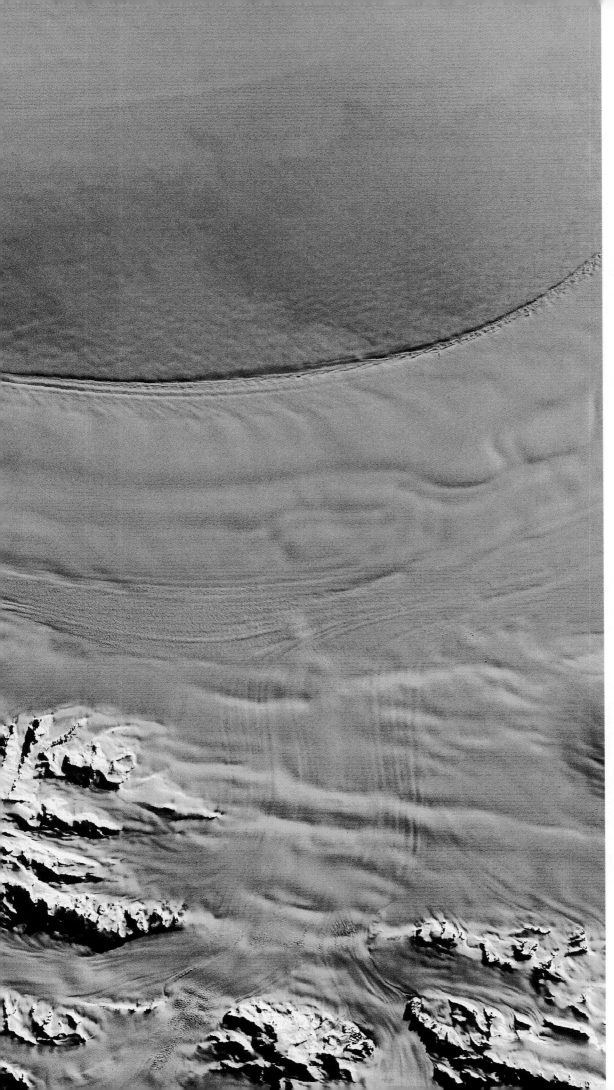

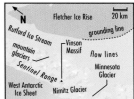

Sentinel Range

The sweeping lines of the Rutford Ice Stream dominate this image in the Ellsworth Mountains. The mountains separate the West Antarctic Ice Sheet (lower left) from the Rutford Ice Stream, which flows (left to right) across the upper half of this Landsat scene. Ice streams are parts of an ice sheet that flow faster than adjacent ice. Vinson Massif, on the crest of the Sentinel Range, at 4897 m (16,160 ft.) is the highest point in Antarctica.

The Fletcher Ice Rise is partially outlined by a well-defined grounding line, which marks the boundary between ice grounded on underlying rock and floating ice. The thickness of ice along the grounding line ranges from 1800 to 2000 m (5,940 to 6,600 ft.), according to measurements obtained with radio-echo sounding instruments.

Several networks of mountain glaciers (center left) flow into the Rutford Ice Stream from the Sentinel Range. Two parallel glaciers, the Nimitz and the Minnesota, carry ice down to the stream from the West Antarctic Ice Sheet. Flow lines visible on the glaciers and ice stream parallel the flow direction.

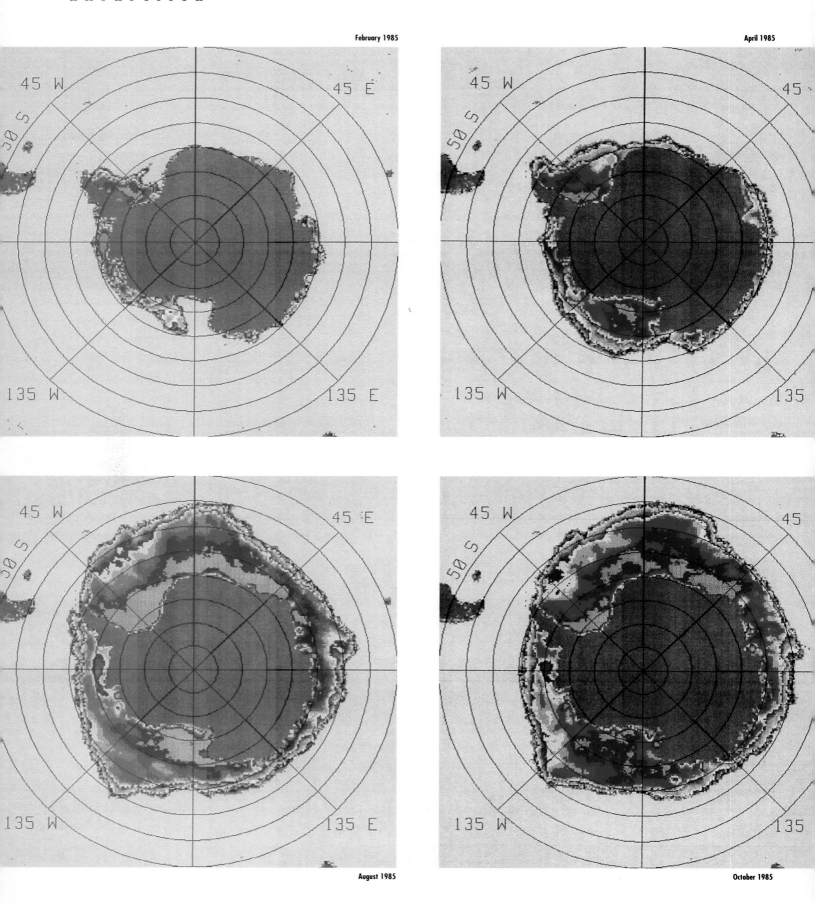

February 1985

April 1985

August 1985

October 1985

June 1985

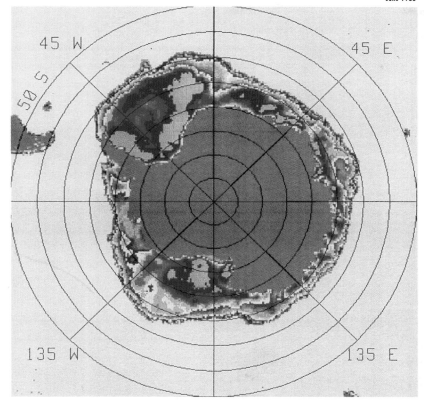

Antarctic Sea Ice

Antarctica's seasonal changes, in these images, form an ice calendar of the continental year. Colors indicate the percentage of the ocean surface covered by sea ice. Violet to pink shows 100 to 76 percent; the brown-to-yellow colors signify a 75 to 44 percent range. Dark green to light blue shows areas with lessening percentages.

Inhibiting heat exchange between the oceans and the atmosphere is one way the fluctuating sea ice (below) influences global climate and environment. It also reflects solar radiation and affects circulation patterns in the ocean and atmosphere. Changes in the distribution and concentration of Antarctic sea ice were measured in 1985 by the

Scanning Multichannel Microwave Radiometer (SMMR, pronounced "simmer") on the Nimbus-7 satellite. The location of South America and the Falkland Islands is indicated on the left side of each image.

The extent of the ice in the coldest months may be influenced by the position of the Circumpolar Current, which flows eastward around the entire continent. In December (left, lower) a polynya, an open area of ocean, forms adjacent to the Ross Ice Shelf. Currents in the Ross Sea and winds off of the shelf may control the polynya's formation and appearance. Such open areas appear in the Antarctic spring as the sea ice begins to melt. The sea ice cover drops to a minimum in the Antarctic summer.

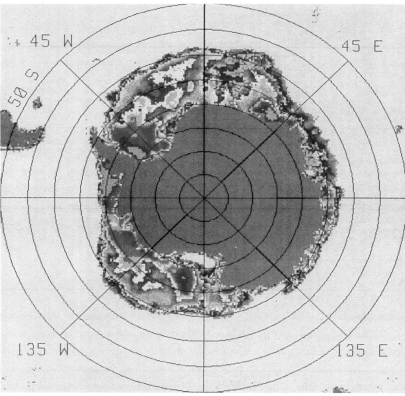

December 1985

Scott and Amundsen Race to the Pole

IN 1911 ROALD AMUNDSEN, A NORWEGIAN EXPLORER, and Robert Falcon Scott, from Great Britain, competed to be the first to set foot on the South Pole. Scott planned to follow the route of Sir Ernest Shackleton who had traveled to within about 155 km (97 mi.) of the pole a few years earlier. Scott began his journey at McMurdo Sound (see opposite page) and landed on the west coast of Ross Island at a promontory he named Cape Evans. Arriving there in January 1911, Scott's party wintered at their camp.

They set out for the pole the following November, proceeding across the Ross Ice Shelf. Unlike Amundsen, who used dogs to pull his sleds, Scott relied on ponies and motor sledges. He planned to send his dogs back before climbing to the Polar Plateau so that they could be used to haul food from the base to restock supply depots along the return route. But the ponies, unsuited to the Antarctic environment, all died before the expedition arrived at a crucial site: the Transantarctic Mountains, which had to be crossed to reach the pole. The motor sledges broke down and could not be repaired.

Without animals or motorized sledges, the team man-hauled all their supplies and equipment (about 200 pounds per man) up the Beardmore Glacier, following Shackleton's route through the mountains. They reached the Polar Plateau by New Year's Day 1912 and proceeded on to finish the remaining 240 km (150 mi.), not knowing that Amundsen had already reached the pole. The arduous trip of nearly 2.5 months not only was a disappointment for the British explorers, but also ended in tragedy. Suffering from frostbite and perhaps scurvy, and lacking adequate provisions, Scott and his team died on the return journey.

The hut at Cape Evans (top) was base for Captain Scott's expedition and remains a memorial to Scott and four others, who perished returning from the South Pole. At the head of the table, Scott (above, center) celebrates his 43rd, and last, birthday.

Amundsen had tried a different approach, landing farther east on the Ross Ice Shelf, thus placing his starting point some 100 km (60 mi.) closer to the pole. He also began his polar journey about 12 days earlier than Scott. After trekking to the southern end of the shelf, he and his men crossed the Transantarctic Mountains by making a steep and difficult climb along a glacier that Amundsen named the Axel Heiberg after a patron of his earlier expeditions. That obstacle crossed, they made their way to the pole, arriving some 34 days ahead of Scott.

Amundsen's entire journey covered unexplored territory. His planning and preparation were based only on general experience and study. He had no direct knowledge of the hazards he would encounter. Today, our understanding of the most remote areas of the Earth is enhanced by the technology of exploration from orbit. Satellite images survey areas difficult to traverse and inhospitable to land explorers, thus aiding science immeasurably. Amundsen himself once predicted that technology would revolutionize Antarctic exploration. He felt that aircraft could eliminate the need for the arduous land journeys. Although disagreeing that air surveys alone would be sufficient, Admiral Richard Byrd, the U.S. explorer after whom Byrd Glacier is named, pioneered the use of aircraft and aerial photography in Antarctica. Establishing his first base, Little America, near Amundsen's camp at the Bay of Whales, Byrd was first to fly over the South Pole. On that flight in November 1929, he compressed Amundsen's three-month journey into 19 hours. His later expeditions utilized ships, planes, and helicopters to gather thousands of photographs charting many areas never before explored.

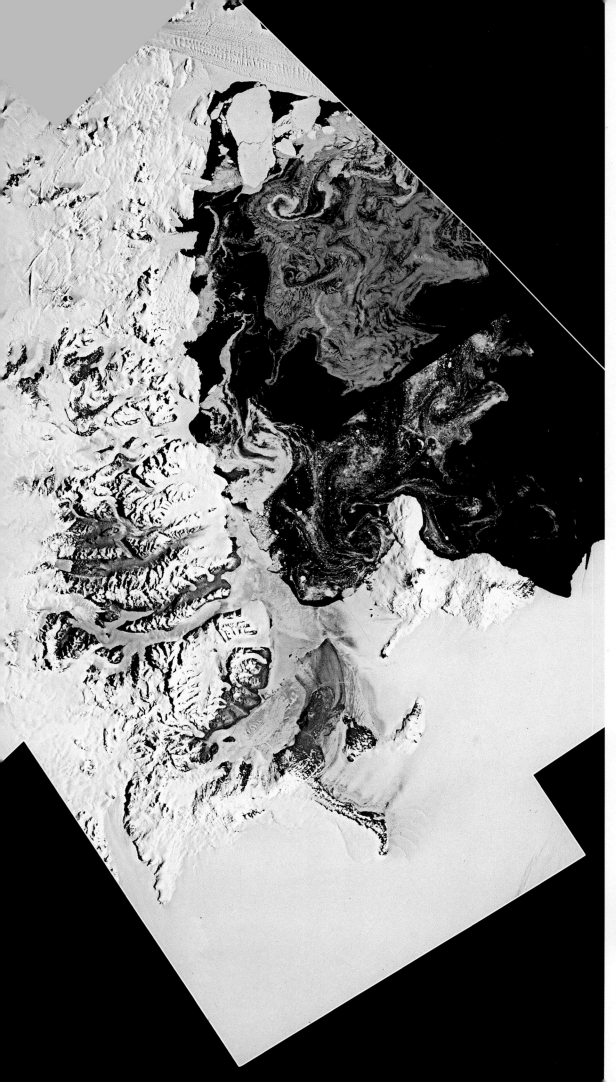

McMurdo Sound

Pack ice forms intricate swirls in McMurdo Sound, off the coast of Victoria Land. Ross Island, at the edge of the Ross Ice Shelf, is made up of three volcanoes. Mount Erebus, the southwestern one, is active. Many explorers have seen its plume on their Antarctic journeys. Along the coast are the Transantarctic Mountains, which divide eastern elevated terrain from low-lying western areas. The rare.dry valleys (center) are virtually clear of snow and contain large glaciers flowing seaward.

Farther to the north are the Allan Hills, the site of hundreds of meteorite finds. Antarctic meteorites are often discovered in blue ice patches such as those around the hills. These patches are areas that have been scoured by the wind to reveal underlying ice. The meteorites may travel along with the ice flow and perhaps concentrate in regions where ice movement is obstructed.

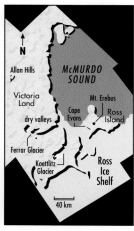

(previous pages)

Bay of Whales

The Bay of Whales, the inlet in the center of the image, marks the site where, in 1911, Roald Amundsen moored his ship, the *Fram*. This was the start of his successful expedition, using dog sleds and skis, to the South Pole. The bay was named by the Antarctic explorer Ernest Shackleton, who saw it filled with whales during his voyage in 1908. At that time Antarctic waters attracted not only whales but also whalers, who nearly wiped out some species.

Amundsen's base camp, called Framheim (home of the *Fram*), was erected on the Ross Ice Shelf east of the bay, about 4 km (2.5 mi.) from the water's edge. The explorers spent several months collecting and storing seal and penguin meat for food. They also made trips inland to place storage depots along the first part of the route to the pole.

After wintering at Framheim, Amundsen first attempted his polar trek in September. Turned back by bad weather, he started again the following month and succeeded. The 99-day round trip covered 2993 km (1,860 mi.).

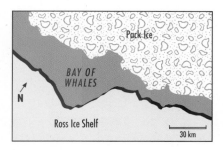

Iceberg Calving

Roald Amundsen poses (left) aboard his ship, the *Fram*, in Oslo harbor. When he landed on the Ross Ice Shelf (above), he set his camp there. A hut was constructed for living quarters, and 16 large tents were erected.

Amundsen thought the camp was firmly grounded over land and that there was little danger from the calving of the ice. But the shelf is floating. In recent years a large section bordering the bay, an iceberg some 30 km (19 mi.) wide, has broken off into the sea.

A sequence of Defense Meteorological Satellite images records the separation of the iceberg from the Ross Ice Shelf in the vicinity of the Bay of Whales. The first scene (opposite, upper), from November 1986, shows the bay intact. Subsequent scenes (October 1987 and December 1988) reveal the breakup of the ice on the edge of the shelf.

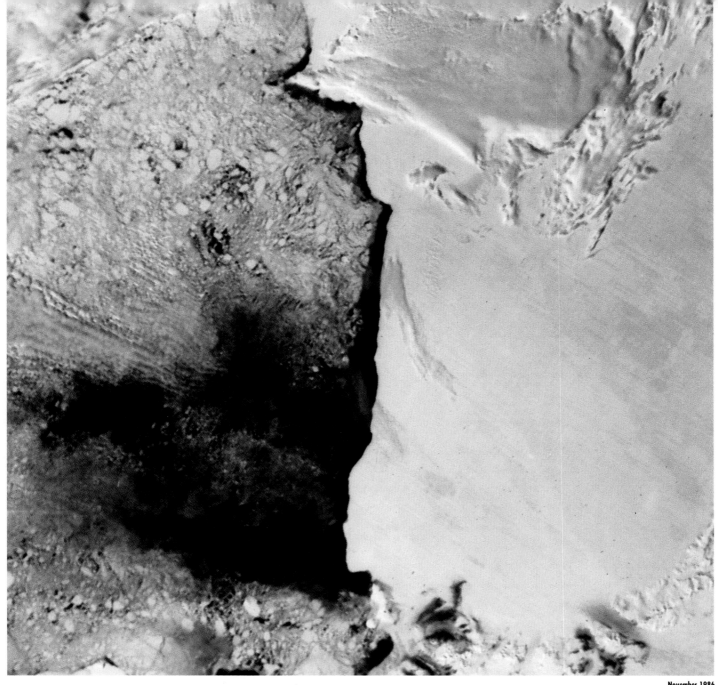

November 1986

October 1987

December 1988

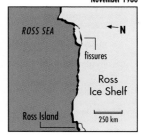

image and photo credits

SOURCES OF IMAGERY

Australian Centre for Remote Sensing
P.O. Box 28
Belconnen, A.C.T. 2616
Australia
(61) (06) 252-4411
fax: (61) (06) 251-6326

Center for Earth and Planetary Studies
National Air and Space Museum
Smithsonian Institution
Washington, D.C. 20560
(202) 357-1424

Central Trading Systems
611 Ryan Plaza Drive, Suite 700
Arlington, TX 76011
(817) 459-0423

Cirrus Technology
116 Perimeter Road
P.O. Box 1126
Nashua, NH 03061-1126
(603) 882-2619

Department of Scientific and Industrial
Research (DSIR)
Bell Road
P.O. Box 31-311
Lower Hutt, New Zealand
(64) (04) 666 919
fax: (64) (04) 690 067

DLR-Forschungszentrum
Post Wessling/OBB
D-8031 Oberpfaffenhofen
Germany
(49) (081 53) 28-0
fax: (49) (081 53) 28-243

Earth Satellite Corporation
6011 Executive Boulevard, Suite 400
Rockville, MD 20852
(301) 231-0660

Environmental Research Institute of
Michigan (ERIM)
P.O. Box 8618
Ann Arbor, MI 48107
(313) 994-1200

EOSAT
4300 Forbes Boulevard
Lanham, MD 20706-9954
(301) 552-0500

EROS Data Center
U.S. Geological Survey
Sioux Falls, SD 57198
(605) 594-6151

Földmérési és Távérzékelési Intézet
1051 Budapest V.
Guszev u. 19
H-1373. Pf. 546, Hungary
(36) (1) 126-480
telex: 22-4964

GLAVKOSMOS
9 Krasnoproletarskaya Street
103030 Moscow, Russia
(7) 258-22-20
telex: 411879 START SU

Goddard DAC
Earth Resources Browse Facility
Code 935
NASA/Goddard Space Flight Center
Greenbelt, MD 20771

IMAX Systems Corporation
38 Isabella Street
Toronto, Ontario
Canada M4Y 1N1
(416) 960-8509
fax: (416) 960-8596

Instituto Nacional de Pesquisas Espaciais
(INPE)
Av. Dos Astronautas, 1758
Caixa Postal 515-12201
São José dos Campos-SP, Brasil
(55) (123) 41.8977
telex: (0123) 3530 INPE BR

Lunar and Planetary Institute
3600 Bay Area Boulevard
Houston, TX 77058-1113
(713) 486-2182

MacDonald Dettwiler
3751 Shell Road
Richmond, British Columbia
Canada V6X 2Z9
(604) 278-3411
fax: (604) 273-9830

Mid-America Remote Sensing Center (MARC)
Murray State University
Murray, KY 42071
(502) 762-2148

NASA/John C. Stennis Space Center
Science and Technology Laboratory
HA-10
Stennis Space Center, MS 39529-6000
(601) 688-1906

NASA/Lyndon B. Johnson Space Center
Flight Science Support Office
Solar System Exploration Division
Mail Code: SN15
Houston, TX 77058
(713) 483-5066

National Geographic Society
17th & M Streets, NW
Washington, D.C. 20036
(202) 857-7000

National Oceanographic and Atmospheric
Administration/National Environmental
Satellite Data and Information Service/
National Climate Data Center
(NOAA/NESDIS/NCDC)
Room 100 Princeton Executive Square
Washington, D.C. 20233
(301) 763-8400

National Remote Sensing Centre, Ltd. (NRSC)
North Gate Road
Farnborough, Hampshire GU14 6TW
United Kingdom
(44) (0252) 541464
fax: (44) (0252) 375016

National Snow and Ice Data Center
Cooperative Institute for Research in
Environmental Sciences (CIRES)
Campus Box 449
University of Colorado
at Boulder
Boulder, CO 80309
(303) 492-5171

National Space Development Agency of
Japan (NASDA)
1401 Ohashi
Hatoyama-Machi
Hikigun, Saitama 350-03, Japan
(81) 3-435-6111
fax: (81) 3-433-0796

National Space Science Data Center (NSSDC)
Code 933.0
NASA/Goddard Space Flight Center
Greenbelt, MD 20771
(301) 286-6695

Natural Systems Analysts, Inc.
1331 Palmetto Avenue, Suite 110
Winter Park, FL 32789
(407) 644-7018

Remote Sensing Applications Center
Department of Land Administration
Jardine House
184 Street George's Tce.
Perth, Western Australia 6000
Australia
(61) (09) 323 1520
fax: (61) (09) 321 8576

Remote Sensing Technology Center of
Japan (RESTEC)
Uni-Roppongi Building
7-15-17, Roppongi
Minato-ku
Tokyo 106 , Japan
(81) 03-403-1761
fax: (81) 03-403-1766

Russian Scientific and Research Institute
of Space Devices
53 Aviamotornaya Street
111024 Moscow, Russia
fax: (7) 273-19-37
telex: 412176 INTEC

Sojuzkarta
45 Volgogradskij Pro.
109125 Moscow, Russia
(7) 177-40-50
telex: 411942 REN SU

Spaceshots, Inc.
11111 Santa Monica Boulevard
Suite 210
Los Angeles, CA 90025
(213) 478-8802

SPOT Image Corporation
1897 Preston White Drive
Reston, VA 22091-4368
(703) 620-2200

TASC
55 Walkers Brook Drive
Reading, MA 01867
(617) 942-2000

Terra-Mar Resource Information
Services, Inc.
1937 Landings Drive
Mountain View, CA 94043-0839
(415) 964-6900

Tokai University Research and Information
Center (TRIC)
2-28-4 Tomigaya, Shibuya-ku
Tokyo 151 Japan
(81) (3033) 481-0611
fax: (81) (3033) 481-0610

U.N. Economic Commission for Africa
P.O. Box 3001
Addis Ababa, Ethiopia
(251) (01) 45 70 00 - 44 72 00

U.S. Geological Survey
Flagstaff Image Processing Facility
Images available from:
U.S. Geological Survey Photo Library
MS 914, Box 25046, Federal Center
Denver, CO 80225
(303) 234-1010

IMAGE AND PHOTO CREDITS

Cover image courtesy of TASC, EOSAT Corporation and CIRRUS Technology
pages 2-3 Landsat MSS frame 10112-08074 ©Earth Satellite Corporation
pages 4-5 Landsat false-color mosaic processed by Chris Peterson, courtesy National Geographic Society
pages 6-7 Landsat TM frame 40149-17440 courtesy ERIM
pages 8-9 Landsat TM scene, courtesy ERIM and Spaceshots, Inc.
page 12 Courtesy of NASA
page 13 Courtesy TRIC/Tokai University Research and Information Center
page 14 Courtesy Defense Audiovisual Agency
page 16 Courtesy International Museum of Photography, Rochester, New York
page 17 Courtesy Deutsches Museum, Munich
page 20 Courtesy Defense AudiovisualAgency
page 21 National Air and Space Museum
page 22 Courtesy CIA
page 23 Robert Aden White via 554th ARTW/SAC
page 24 Space Shuttle Mission STS-29, frame 71-046
page 25 NASA photo
page 26 Salton Sea visible—Apollo 9 AS9-23-3558, NASA photo/Salton Sea infrared—Apollo 9 AS9-26A-3800A, NASA photo/Salton Sea radar—Seasat radar image, Rev. 1140, processed by Jet Propulsion Laboratory
page 27 Landsat TM frame 52051-15114X0, processed by CEPS/National Air and Space Museum
page 29 (top) Landsat TM, frame 50150-33009129110, processed by CEPS/National Air and Space Museum
page 29 (bottom) Landsat TM courtesy JPL
page 29 Excavation at Ubar (bottom right)

page 31 NOAA AVHRR mosaic, courtesy ERIM and National Geographic Society
page 34 Rift Valley—©Robert Caputo
page 35 Landsat MSS mosaic, courtesy United Nations Economic Commission for Africa
page 36 Space Shuttle Mission 61-C, frame 35-055
page 37 Djibouti—©Chris Jones/AllStock (top)
page 37 Space Shuttle Mission 61-B, frame 43-071 (bottom)
page 38 Mali—©Betty Press/Picture Group
page 39 U.S. Air Force Defense Meteorological Satellite Program image, courtesy National Snow and Ice Data Center/CIRES
pages 40, 41 (top) Landsat MSS frame 3810-95645, courtesy of P. A. Jacobberger, CEPS/National Air and Space Museum
page 41 (bottom) Landsat MSS frame 9350-10082, courtesy of P. A. Jacobberger, CEPS/National Air and Space Museum
pages 42-43 Landsat MSS frame 1110-07561, ©Earth Satellite Corporation
page 43 Dunes—©Georg Gerster (bottom)
page 44 (top) Image courtesy U.S. Geological Survey. Shuttle Imaging Radar-A, data take 28. Principal Investigator Charles Elachi, Jet Propulsion Laboratory
page 44 (bottom) Space Shuttle Mission 51-I, frame 40-010
page 45 Space Shuttle Mission 51-I, frame 33-052
page 46 Space Shuttle Mission 61-A, frame 43-023
page 47 Beetle—©William J. Hamilton III
pages 48-49 Space Shuttle Mission 41-D, frame 39-099
page 50 Abu Simbel—©Robert Caputo
page 51 Landsat MSS mosaic, ©Earth Satellite Corporation
page 52 Space Shuttle Mission 41-G, Large Format Camera frame 1315, courtesy ITEK Optical Systems
page 53 Space Shuttle Mission STS-26, frame 34-074
pages 54-55 ©1990 CNES, provided by SPOT Image Corporation
page 56 Space Shuttle Mission 61-A, frame 43-029
page 57 Landsat TM frame 5012-60832, courtesy Goddard DAC, Earth Resources Browse Facility (top)
page 57 Congo River—©Robert Caputo (bottom)
page 58 Madagascar—©Frans Lanting/Minden Pictures
page 59 Space Shuttle Mission 51-I, frame 39-042
page 61 NOAA AVHRR mosaic, ©Earth Satellite Corporation
page 64 Mount Everest—©William Thompson
page 65 Three-dimensional perspective images courtesy TRIC/Tokai University Research and Information Center
page 66 ESA Spacelab-1 Metric Camera frame 01-0013-01, Space Shuttle Mission STS-9, courtesy DLR/German Aerospace Research Establishment (top)
page 66 Everest—©Barry Bishop/National Geographic Society, 1963 (bottom)
page 67 Space Shuttle Mission STS-27, frame 33-079
page 68 Landsat Mosaic, ©NASDA
page 69 Old Volcano Mt. Sakura-jima—©T.A. Jaggar/National Geographic Society, 1924 (top)
page 69 ©1990 CNES, provided by SPOT Image Corporation (bottom)
pages 70-71 Landsat/topographic data merge, ©RESTEC/Nihon University, 1990
page 72 Mt. Tolbachik Volcano—©Vadim Gippenreiter/John Calmann & King Ltd. (top)

page 72 Shuttle Imaging Radar-A image, data take 35-36 acquired through the National Space Science Data Center. Principal Investigator, Charles Elachi, Jet Propulsion Laboratory (bottom)
page 73 Space Shuttle Mission 61-A, frame 45-096
page 74 Yangtze—©How Man Wong
page 75 Landsat TM satellite image, reproduced with permission of the Earth Observation Satellite Co. (EOSAT)
page 76 Space Shuttle Mission 61-A, frame 43-007
page 77 Landsat MSS frame 30614-02250, ©Earth Satellite Corporation
page 78 Space Shuttle Mission 61-B, frame 50-007
page 79 NOAA AVHRR imagery, courtesy R. M. Carey, National Environmental Data and Information Service, NOAA
page 80 Forbidden City—©Dean Conger/Lee & Dean Productions
page 81 © 1992 CNES, provided by SPOT Image Corporation
page 82 Grand Canal—©Charles O. Hyman
page 82 Shuttle Imaging Radar-A image, data take 35-36, acquired through the National Space Science Data Center. Principal Investigator, Charles Elachi, Jet Propulsion Laboratory (bottom)
page 83 Landsat TM frame 51173-01485, courtesy ERIM
pages 84-85 Landsat MSS frame 10073-04181, ©Earth Satellite Corporation
page 86 Aral Sea—©David Turnley/Black Star
page 87 Resource-01 MSU-SK image, courtesy GLAVKOSMOS and the Russian Scientific and Research Institute of Space Devices
page 88 MOS-1 image, ©NASDA/RESTEC, 1987
page 89 Tokyo—©Karen Kasmauski
page 90 Hong Kong—©Jodi Cobb/National Geographic Society, 1991 (top)
page 90 ©1990 CNES, provided by SPOT Image Corporation (bottom)
page 91 Landsat MSS frame 82361-02525500, courtesy P.A. Jacobberger, CEPS/National Air and Space Museum
page 92 ©1991 CNES, provided by SPOT Image Corporation
page 93 Landsat MSS frame 10116-05034, ©Earth Satellite Corporation
page 94 Mosaic courtesy EROS Data Center, National Mapping Division, U.S. Geological Survey; data courtesy NOAA
page 98 Oman Green Mountains—©Lynn Abercrombie
page 99 Space Shuttle Mission STS-28, frame 152-083
page 100 Space Shuttle Mission STS-27, frame 152-006
page 101 Space Shuttle Mission 41-G, frame 37-105
pages 102-3 IMAX camera scene, ©IMAX Systems Corporation
page 103 Footsteps of Moses/Sinai—©Nathan Benn/Woodfin Camp (top)
page 104 Space Shuttle Mission STS-28, frame 96-065
page 105 Landsat MSS mosaic, ©Earth Satellite Corporation
page 106 Landsat TM frame 51249-07093, courtesy EROS Data Center
page 107 Mt. Ararat—©Harry Naltchayan (top)
page 107 Space Shuttle Imaging Radar-A image, data take 35-36, acquired through the National Space Science Data Center. Principal Investigator Charles Elachi, Jet Propulsion Laboratory (bottom)
page 108 Space Shuttle Mission STS-37, frame 152-091
page 109 Space Shuttle Imaging Radar-A image, data take 37A, acquired through the National Space Science Data Center

image and photo credits

Principal Investigator, Charles Elachi, Jet Propulsion Laboratory (top)
page 109 Oil Fires Kuwait—©Laurent Chamussy/ SIPA Press (bottom)
page 110 Rub al Khali—©Robert Azzi/Woodfin Camp
page 111 Landsat MSS mosaic, ©Earth Satellite Corporation
pages 112-13 Landsat TM frame 50206-0647303, ©Earth Satellite Corporation
page 114 Space Shuttle Mission STS-34, frame 78-078
page 115 Landsat MSS frame 20199-06150, ©Earth Satellite Corporation
page 116 NOAA AVHRR mosaic, image supplied by NRSC, Ltd., with the agreement of NRSC
page 120 Land's End—©Michael Reagan
page 121 Space Shuttle Mission 61-A, frame 201-124
page 122 Crete—©Ira Block (top)
page 122 Space Shuttle Mission STS-28, frame 96-061 (bottom)
page 123 Apollo-Soyuz Test Project scene, frame AST-13-834
page 124 Space Shuttle Mission STS-28, frame 74-001
page 125 Astrakhan—©Charles O. Hyman (top)
page 125 Space Shuttle Mission STS-39, frame 151-007 (bottom)
page 126 Seine, Paris—©James L. Stanfield/ National Geographic Society, 1989
page 127 ©1990 CNES, provided by SPOT Image Corporation
page 128 Landsat MSS mosaic, image supplied by NRSC, Ltd., with the agreement of NRSC
page 129 Landsat MSS frame 1426-12070, courtesy R. S. Williams, Jr., U.S. Geological Survey
pages 130-31 NOAA AVHRR image, courtesy DLR/German Aerospace Research Establishment
page 132 Marseille—©Julien Neiman/Susan Griggs Agency
page 133 Space Shuttle Mission STS-28, frame 78-064
page 134 Skylab 2, frame SL2-5-370
page 135 Landsat TM frame 50147-18305, courtesy Goddard DAC, Earth Resources Browse Facility (top)
page 135 Barcelona—©Robert Frerck/Odyssey Productions (bottom)
page 136 St. Petersburg—©Steve Raymer (top)
page 136 KFA-1000 Film 49 Frame 18108, Sojuzkarta and Central Trading Systems (bottom)
page 137 U.S. Air Force Defense Meteorological Satellite Program mosaic, courtesy National Snow and Ice Data Center, CIRES
page 138 ©1990 CNES, provided by SPOT Image Corporation
page 139 Berlin Wall—©David Alan Harvey
pages 140-41 Landsat TM frame 20102-484295, courtesy ERIM
page 142 Piazza San Marco, Venice—©Michael S. Yamashita (top)
page 142 Space Shuttle Mission 41-G, Large Format Camera frame 1284, acquired through the National Space Science Data Center. Principal Investigator Bernard H. Mollberg (bottom)
page 143 ©1992 CNES, provided by SPOT Image Corporation
page 144 Pollution Czechoslovakia—©Shepard Sherbell/Saba
page 145 Landsat TM frame 42940-09321, data courtesy Barrett Rock, University of New Hampshire; processed by Michael J. Tuttle, CEPS/National Air and Space Museum
page 147 Landsat TM frame, 189-27/4, courtesy Földmérési És Tavérzékelési Intézet
page 148 Landsat TM/SPOT merge, ©1990 CNES, provided by SPOT Image Corporation (top)
page 148 Chernobyl—©Steve Raymer (bottom)
page 149 NOAA AVHRR image, courtesy DLR/German Aerospace Research Establishment
page 151 NOAA AVHRR mosaic, processed and supplied by Terra-Mar Resource Information Services, Inc.
page 154 Grand Canyon—©Jeff Gnass
page 155 Space Shuttle Mission STS-36, frame 151-002
page 156 Space Shuttle Mission 39, frame 151-087
page 157 Space Shuttle Mission 51-B, frame 147-030
page 158 Space Shuttle Mission STS-36, frame 51-225
page 159 Seasat image, Rev. 759, NOAA/NESDIS
page 160 Everglades—©Farrell Grehan (top)
page 160 Space Shuttle Mission 51-C, frame 143-032 (bottom)
page 161 Space Shuttle Mission 51-C, frame 143-027
page 162 Space Shuttle Mission 51-B, frame 146-122
page 163 HCMM scene, 054-08170-3, night IR, acquired through the National Space Science Data Center; Principal Investigator William L. Barnes
pages 164-65 Space Shuttle Mission 61-A, frame 201-075
page 167 Space Shuttle Mission 51-F, frame 38-045
pages 168-69 Landsat MSS scenes, courtesy ERIM
page 169 Mt. St. Helens before & after eruption—©Michael Lawton (bottom 2)
page 170 Ellesmere Island—©Jim Brandenburg/Minden Pictures (top)
page 170 U.S. Air Force Defense Meteorological Satellite Program mosaic, courtesy National Snow and Ice Data Center/CIRES (bottom)
page 171 Space Shuttle Mission STS-36, frame 75-062
page 172 Chicago—©Lynn Johnson/Black Star
page 173 ©1990 CNES, provided by SPOT Image Corporation
page 174 New York—©Michael Reagan
page 175 U.S. Air Force Defense Meteorological Satellite Program mosaic, courtesy National Snow and Ice Data Center/CIRES
page 176 ©1991 CNES, provided by SPOT Image Corporation
page 177 St. Louis Arch—©Fred J. Maroon (top)
page 177 Space Shuttle Mission 51-B, frame 146-158 (bottom)
pages 178-79 Landsat TM scenes, processed by CEPS/National Air and Space Museum
page 180 San Andreas Fault—©Barrie Rokeach (top)
page 180 Space Shuttle Mission 51-J, frame 39-056 (bottom)
page 181 Space Shuttle Mission 61-A, frame 44-036
page 182 Vancouver—©Chris Johns/AllStock
page 183 Landsat TM scene, courtesy MacDonald Dettwiler, enhanced by Advanced Satellite Production
page 185 NOAA AVHRR mosaic, courtesy ERIM and National Geographic Society, additional processing by CEPS/National Air and Space Museum
page 188 Popocatépetl —©Mark Godfrey/ National Geographic Society, 1981
page 189 Space Shuttle Mission STS-30, frame 151-158
page 190 Space Shuttle Mission STS-31, frame 91-068
page 191 Space Shuttle Mission 61-C, frame 31-037 (top)
page 191 El Chichon—©Ken Garrett/Woodfin Camp (bottom)
page 192 Space Shuttle Mission 61-C, frame 37-076
page 193 Space Shuttle Mission 61-C, frame 31-048
page 194 Tikal—©Bill Garrett
page 195 Landsat TM mosaic, courtesy Thomas Sever, NASA/Stennis Space Center
page 196 Landsat TM frame 42304-15552, ©Earth Satellite Corporation
page 197 Space Shuttle Imaging Radar-A image, data take 37, acquired through the National Space Science Data Center. Principal Investigator, Charles Elachi, Jet Propulsion Laboratory
page 198 Landsat TM frame 50707-15211, courtesy Steven A. Sader, University of Maine and NASA/Stennis Space Center (top)
page 198 Monteverde Cloud Forest—©Tom Till (bottom)
page 199 Landsat MSS frame 21455-14205, courtesy ERIM
page 200 Puerto Rico—©Farrell Grehan
page 201 Space Shuttle Mission STS-29, frame 90-012
pages 202-3 Space Shuttle Mission STS-29, frame 90-087
page 204 Landsat TM image courtesy Natural Systems Analysts, Inc. (top)
page 204 Bimini—©Stephan Frink/The Waterhouse (bottom)
page 205 Landsat MSS frame 21678-14102, ©Earth Satellite Corporation
page 206 Kingston Port—©W. Bertsch/Bruce Coleman, 1992 (top)
page 206 Space Shuttle Mission STS-36, frame 152-130 (bottom)
page 209 NOAA AVHRR mosaic courtesy EROS Data Center, National Mapping Division, U.S. Geological Survey; data courtesy NOAA
page 212 Rio slums—©Stephanie Maze
page 213 Landsat MSS frame 20391-14275, courtesy ERIM
page 214 Cuzco—©Loren McIntyre
page 215 ©1991 CNES, provided by SPOT Image Corporation
page 216 Landsat MSS frame 2228-84123, courtesy ERIM
page 217 Landsat TM path/row 221/071, courtesy INPE
pages 218-19 Landsat MSS frame 77212-13081, ©Earth Satellite Corporation
page 220 Rio harbor—©Stephanie Maze
page 221 Space Shuttle Mission 61-C, frame 34-020
page 222 Llamas—©Loren McIntyre
page 223 Satellite imagery/topographic data merge, courtesy TRIC/Tokai University Research and Information Center
page 224 Landsat MSS frame 1244-14065, ©Earth Satellite Corporation
page 225 Space Shuttle Mission STS-37, frame 152-101
page 226 Space Shuttle Mission STS-26, frame 40-060
page 227 Lake Titicaca—©Courtney Milne
pages 228-29 Space Shuttle Mission STS-26, frame 40-055
page 230 Iguazu Falls ©James P. Blair/ National Geographic Society, 1992
page 231 Landsat TM frame 51099-12573, ©Earth Satellite Corporation
page 232 Landsat TM frame 58907-91339, ©Earth Satellite Corporation
page 233 Landsat TM frame 58821-41335, ©Earth Satellite Corporation
pages 234-35 Landsat TM frame 58907-41321, ©Earth Satellite Corporation
page 237 Space Shuttle Mission 51-J, frame 144-075
page 238 Landsat TM frame 85083-143551, courtesy Luis Bartolucci, Mid-America Remote Sensing Center (top)
page 238 Nazca Line—©Bob Sacha (bottom)
page 239 ©1990 CNES, provided by SPOT Image Corporation
page 240 NOAA AVHRR mosaic, provided by the Remote Sensing Applications Centre, Australian Department of Land Administration
page 244 Barrier Reef—©Ron Taylor/Bruce Coleman, 1992
page 245 Space Shuttle Mission 41-G, Large Format Camera frame 1752, acquired through the National Space Science Data Center; Principal Investigator Bernard H. Mollberg
page 246 Space Shuttle Mission STS-7, frame 19-923 (top)
page 246 Shark Bay—©Jan Taylor/Bruce Coleman, 1992 (bottom)
page 247 Space Shuttle Mission 61-A, frame 49-039
page 248 Space Shuttle Mission STS-9, frame 35-1622
page 249 Space Shuttle Mission 41-G, Large Format Camera Frame 1706, acquired through the National Space Science Data Center; Principal Investigator Bernard H. Mollberg
pages 250-51 Space Shuttle Mission 41-G, Large Format Camera frame 1696, acquired through the National Space Science Data Center; Principal Investigator Bernard H. Mollberg
page 252 Desert Theme—©A. Fox/Auscape
page 253 Space Shuttle Mission 41-D, frame 41-016
page 254 Landsat MSS, path/row 117/73, processed by Australia Centre for Remote Sensing (top)
page 254 Great Sandy Desert—©Sam Abell/National Geographic Society, 1991 (right)
page 254 Space Shuttle Mission 41-D, frame 41-047 (bottom)
page 255 Space Shuttle Mission 51-A, frame 46-034
page 256 Gibson Desert—©Jean-Paul Ferrero/Auscape (top)
page 256 Space Shuttle Mission 41-D, frame 42-045 (bottom)
page 257 Landsat MSS frame 10207-00165, ©Earth Satellite Corporation
page 258 Tasman Glacier—©NickDevore/ Bruce Coleman, 1992
page 259 Landsat MSS mosaic, processed by Department of Scientific and Industrial Research, DSIR
page 260 Space Shuttle Mission 61-A, frame 51-015
page 261 Space Shuttle Mission STS-9, frame 40-2572
page 263 Space Shuttle Mission 61-A, frame 42-084
page 264 Bora Bora—©Nick Devore/Bruce Coleman, 1992
page 265 Space Shuttle Mission 51-B, frame 146-028
page 266 Space Shuttle Mission STS-8, frame 32-0748
page 267 Space Shuttle Mission 41-D, frame 33-046
pages 268-69 ©1990 CNES, provided by SPOT Image Corporation
page 270 Space Shuttle Mission 41-D, frame 44-096
page 271 Space Shuttle Mission 41-D, frame 32-026
page 272 Easter Island heads ©James Balog/Black Star (top)
page 272 Space Shuttle Mission 51-I, frame 45-054 (bottom)
page 273 Space Shuttle Mission STS-26, frame 43-056
page 274 Space Shuttle Mission 61-A, frame 50-057
page 275 Kilauea Volcano ©James Sugar/ Black Star
page 276 NOAA AVHRR mosaic, image supplied by NRSC, Ltd., with the agreement of NRSC
page 280 Mt. Erebus—©Colin Monteath/ Hedgehog House, N.Z.
page 281 Landsat MSS frame 1542-18435, courtesy B. K. Lucchitta and the Flagstaff Image Processing Facility, U.S. Geological Survey
page 282 Landsat MSS frame 1128-20275, courtesy B. K. Lucchitta and the Flagstaff Image Processing Facility, U.S. Geological Survey
page 283 Landsat RBV frame 30928-20440, courtesy R. S. Williams, Jr., and J. Ferrigno, U.S. Geological Survey
page 284 Landsat MSS frame 1170-12260, courtesy of B. K. Lucchitta and the Flagstaff Image Processing Facility, U.S. Geological Survey
page 285 Tabular Iceberg ©Wolfgang Kaehler
pages 286-87 Landsat MSS frame 1560-11492, courtesy B. K. Lucchitta and the Flagstaff Image Processing Facility, U.S. Geological Survey
pages 288-89 SMMR imagery courtesy H. Jay Zwally, NASA/Goddard Space Flight Center
page 289 Floating Ice—©Wolfgang Kaehler
page 290 Scott's hut—©Colin Monteath/Hedgehog House, N.Z.(top)
page 290 Scott's birthday—©Scott Album/Alexander Turnbull Library, Wellington, N. Z. F11331 1/2 page 272 Space Shuttle Mission 51-I, frame 45-054 (bottom)
page 291 Landsat mosaic, courtesy B. K. Lucchitta and the Flagstaff Image Processing Facility, U. S. Geological Survey
pages 292-93 Landsat MSS frame 1491-17181, courtesy B. K. Lucchitta and the Flagstaff Image Processing Facility, U. S. Geological Survey
page 294 Ross Ice—©Colin Monteath/Hedgehog House Shelf (top)
page 294 Amundsen—©Norsk Folkemuseum (bottom)
page 295 U.S. Air Force Defense Meteorological Satellite Program imagery, visible band, 14 Oct 1987, 20 Nov 1986, 4 Dec 1988; courtesy National Snow and Ice Data Center/CIRES
back flap—Carolyn Russo

MAP NOTES

The maps at the beginning of each chapter are presented in the following projections and approximate scales:

Africa: Mercator, 1:28,000,000
Asia: Hammer Azimuthal, 1:33,000,000
Middle East: Lambert Conformal Conic, 1:11,000,000
Europe: Lambert Conformal Conic, 1:18,000,000
North America: Chamberlin Trimetric, 1:24,000,000
Middle America: Lambert Conformal Conic, 1:13,000,000
South America: Lambert Azimuthal, 1:27,000,000
Oceania: Mercator, 1:60,000,000
Antarctica: Polar Stereographic, 1:22,000,000.

Boundaries and sovereignty status on all maps presented in this book are not necessarily those recognized by the U. S. Government.

ALMAZ, Russia

Operational Dates: March 31, 1991 to October 17, 1992
Coverage: 78°N to 78°S
Altitude: 300 km

Sensor	Wavelength	Resolution
Synthetic Aperture Radar (SAR)	10 cm	15 m
Radiometric Scanner (RMS)	5 cm, .8 cm, IR	10-30 km

The Almaz satellite used synthetic aperture radar to obtain information about land and sea surfaces. The data can be used to create photographic products or maps. The experimental radiometric scanner on board scans a 600-km-wide swath at 10- to 30-km resolution.

SAR digital tapes and photographic products available from:
Hughes STX Satellite Mapping Services
4400 Forbes Boulevard
Lanham, MD 20706 USA
(301) 794-5020
fax: (301) 306-0963

Apollo-Soyuz Test Project (ASTP), U.S. and U.S.S.R.

Operational Dates: July 15 to 24, 1975
Coverage: Selected sites between 52°N and 52°S
Altitude: 176 to 232 km
Cameras: 70 mm bracket-mounted Hasselblad
　　　　　70 mm hand-held Hasselblad
　　　　　35 mm Nikon

The ASTP mission marked the first docking of spacecraft from two separate nations. ASTP photography was conducted as part of the Earth Observations and Photography Experiment. More than 1,400 photographs were collected of sites selected to include coverage of geologic, desert, oceanographic, hydrologic and meteorological features. Visual observations by the astronauts were also recorded because of the human eye's ability to distinguish colors and to analyze what is seen instantaneously.

Photographic products available from:
U.S. Geological Survey
EROS Data Center
Customer Services
Sioux Falls, SD 57198 USA
(605) 594-6151
fax: (605) 594-6589

Advanced Very High Resolution Radiometer (AVHRR), U.S.

Mission	Dates
TIROS-N	10/78 to 2/81
NOAA-6	6/79 to 3/87
NOAA-7	6/81 to 6/86
NOAA-8	3/83 to 1/86
NOAA-9	12/84 to present
NOAA-10	9/86 to present
NOAA-11	9/88 to present
NOAA-12	5/91 to present

Band	Wavelength (μm)
1	.58-.68
2	.72-1.0
3	3.55-3.93
4	10.3-11.3
5	11.4-12.4

The AVHRR instrument has flown on eight satellites, all launched into a low polar orbit. Four are still active, but two of these, NOAA-9 and 10, are in standby operation and used only as possible backups. AVHRR bands 1 and 2 are used to show the location of snow, ice, and coastlines. The other bands are in the infrared and are used to discern surface temperature and gather images at night. AVHRR images are also used by meteorologists to observe cloud systems as they move about in the atmosphere. Every AVHRR band has a resolution of about one kilometer. Band 5 data is only available from NOAA-11 at present.

Information and products available from:
NOAA/NESDIS/NCDC
Satellite Data Services Division
5627 Allentown Road
Camp Springs, MD 20746 USA
(301) 763-8400
fax: (301) 763-8443

Camera Systems, Russia
(including former U.S.S.R. systems)

Manned Missions:

Camera	Resolution (m)	Film Sensitivity
MKF-6 (6-camera array, each with different filter)	20	6 Bands (μm) .46-.50, .50-.56 .58-.62, .64-.68 .70-.74, .78-.86
KATE-140	60	panchromatic (B&W)

These camera systems have flown on various Sojuz and Mir missions. The MKF-6 was designed to test multi-spectral photography from space. The KATE-140 was built for orbital mapping.

Unmanned Missions:

Camera	Resolution (m)	Film Sensitivity
KFA-1000 (2 emulsion film)	5 10	panchromatic (B&W) color—2 bands: .57-.67, .67-.81
MK-4 (4 film bases)	6	4 bands varying between .4-.9μm
KATE-200 (3 camera array)	15-20	3 bands (μm): .51-.60, .60-.70, .70-.85

These camera systems have been placed on board various unmanned satellites including those of the Cosmos series. The film is exposed and returned to Earth in re-entry capsules. The cameras provide some of the highest spatial resolution photographs commercially available, but they do not have the spectral resolution of many digital systems.

Photographic products available from:
Central Trading Systems
611 Ryan Plaza Drive, Suite 700
Arlington, TX 76011 USA
(817) 459-0423
fax: (817) 459-0721

Sojuzkarta
45 Volgoradskij pr.
109125 Moscow, Russia
Telex: 411942 REN SU

Defense Meteorological Satellite Program (DMSP), U.S.

Operational Dates: 1971 to present
Coverage: worldwide
Altitude: 833 km
Orbit: near-polar
Primary Sensor: Operational Linescan System (OLS)

The United States Air Force, using a series of DMSP satellites, has provided up-to-date weather forecasting for the U.S. military since 1971. The applications of the OLS include the monitoring of cloud cover, cloud-top temperatures, and ocean-surface temperatures. Band 1 is used to record the visible part of the spectrum at night, with moonlight, and is capable of detecting artificial lighting from cities, large fires, and gas flares from oil production.

Photographic products available from:
NSIDC/DMSP Archive
CIRES, Campus Box 449
University of Colorado
Boulder, CO 80309 USA
(303) 492-2378
fax: (303) 492-2968

Earth Remote Sensing Satellite (ERS-1), European Space Agency

Operational Dates: July 16, 1991 to present
Altitude: 780 km
Orbit: near-polar

Sensor	Resolution
Synthetic Aperture Radar (SAR)	30 m
Scanning Radiometer	1 km

The Europe Remote Sensing satellite transmits and detects microwave radiation to investigate climate and surface features. The SAR is a radar instrument that uses active microwave signals to produce images of the surface. A wind scatterometer measures wind velocity at the sea surface. Also on board are a radar altimeter that determines surface elevation and a scanning radiometer that measures temperature and water vapor. Another instrument, the Laser Retro-Reflector, allows very accurate measurement of the satellite's orbit.

Digital tapes and photographic products available from:
Eurimage
Viae E. D'Onofria 212
00511 Rome, Italy
(39) (6) 406941
fax: (39) (6) 40694232

Gemini III-XII, U.S.

Mission	Operational Dates	Number of Usable Photographs
Gemini III	March 23, 1965	7
Gemini IV	June 3 to 7, 1965	100
Gemini V	Aug. 21 to 29, 1965	175
Gemini VI-A	Dec. 15 to 16, 1965	60
Gemini VII	Dec. 4 to 8, 1965	250
Gemini VIII	March 16 to 17, 1966	None
Gemini IX-A	June 3 to 6, 1966	160
Gemini X	July 18 to 21, 1966	75
Gemini XI	Sept. 12 to 15, 1966	102
Gemini XII	Nov. 11 to 15, 1966	160

Altitude: 161-322 km, Gemini XI 280-1365 km
Cameras: 70 mm Hasselblad 500C — Gemini III - IX-A
70 mm Maurer Space Camera — Gemini IX-A - XII
Hasselblad Super Wide Angle — Gemini IX-A - XII
Coverage: Selected sites between 33°N and 33°S Latitude

Gemini was the second U.S. manned space program. The Gemini spacecraft carried a crew of two and was designed to give more flight control to the astronauts. The primary goals of the program were testing and practicing rendezvous and docking of spacecraft, extravehicular walks, and long-duration missions. These techniques would be needed for the Apollo expeditions to the moon.

The Gemini missions proved that high-quality images of geologic features could be recorded from space and that remote regions of the Earth could be mapped. This created interest in more space-based photography and eventually led to the launching of a satellite to provide continuous monitoring of Earth resources.

Photographic products available from:
U.S. Geological Survey
EROS Data Center
Customer Services
Sioux Falls, SD 57198 USA
(605) 594-6151
fax: (605) 594-6589

Heat Capacity Mapping Mission (HCMM), U.S.

Operational Dates: April 26, 1978 to Sept. 30, 1980
Coverage: United States, Europe, Northwest Africa, Eastern Australia
Altitude: 620 km
540 km (Feb. 1980 to Sept. 1980)
Sensor: Heat Capacity Mapping Radiometer (HCMR)

Channel	Wavelength (μm)	Resolution (nadir)
Visible	0.55-1.1	0.5 km x 0.5 km
Thermal	10.5-12.5	0.6 km x 0.6 km

The sensor on board the HCMM satellite was used to measure the daily cycle of temperature change at the Earth's surface. Measurements of the same area were taken 12 or 36 hours apart, producing day and night coverage. The difference in day and night surface temperatures is an indication of a property called thermal inertia, or resistance of a material to temperature change. Thermal inertia, along with other thermal data and images from HCMM's visible channel, may provide information about a variety of surface properties. These include surface composition, geologic structures, water circulation and moisture content. HCMM data also have applications in detecting local climate features, such as urban heat islands, and in measuring temperature differences across bodies of water to monitor thermal pollution, and warm currents and eddies.

Data and Imagery Archive:
National Space Science Data Center
Code 933.4
National Aeronautics and Space Administration
Goddard Space Flight Center
Greenbelt, MD 20771 USA
(301) 286-6695
fax: (301) 286-4952

Indian Remote Sensing Satellite 1A, 1B (IRS-1A, 1B), India

Operational Dates: March 1988 to present (1A)
August 1991 to present (1B)
Coverage: India and nearby South Asian countries
Altitude: 904 km
Orbit: near-polar
Sensors: 3 Linear Imaging Self-scanned Sensors (LISS)

	Band	Wavelength (μm)	Resolution (m)
LISS-1	1	.45-.52 (blue)	72.5
	2	.52-.59 (green)	72.5
	3	.62-.68 (red)	72.5
	4	.77-.86 (near IR)	72.5
LISS-2, A,B	1	.45-.52 (blue)	36.25
	2	.52-.59 (green)	36.25
	3	.62-.68 (red)	36.25
	4	.77-.86 (near IR)	36.25

The IRS satellites were launched aboard Vostok boosters from Khazhakstan in the former Soviet Union. The satellites were developed to help India manage its Earth resources for such purposes as forestry, agriculture, the location of minerals, and urban planning. At present only one receiving station at Shadnagar, India is operating. The data are processed at the remote sensing facility at Balanagar.

Data and imagery available from:
NRSA Data Centre
National Remote Sensing Agency
Balanagar Hyderabad - 500 037
Andhra Pradesh, India
Telex: 0425-6522

Earth Resources Satellite-1 (JERS-1), Japan

Operational Dates: Feb. 1992 to present
Altitude: 570 km
Orbit: near-polar

Sensor	Wavelength	Resolution
Visible and Near-Infrared Radiometer (VNR)	8 bands (5 visible, 3 IR)	20 m
Synthetic Aperture Radar (SAR)	24 cm	20 m

The Japanese Earth Resources Satellite (ERS-1 or JERS-1) uses two main instruments to look at the Earth. The VNR is able to scan the surface in eight bands, three in the infrared. This capability, combined with its high resolution, may be useful for geologic mapping. The SAR produces radar images of the surface.

Digital tapes and photographic products available from:
Remote Sensing Technology Center (RESTEC)
Uni-Roppongi Building
7-15-17 Roppongi, Minato-ku
Tokyo 106, Japan
(81) (03) 403-1761
fax: (81) (03) 403-1766

Landsat, U.S.

Operational Dates: July 1972 to Jan. 1978 (Landsat 1)
Jan. 1975 to Feb. 1982 (Landsat 2)
March 1978 to March 1983 (Landsat 3)
July 1982 to present (Landsat 4)
March 1984 to present (Landsat 5)

Coverage: 81°N - 81°S
Altitude: 913 km (1,2,3)
705 km (4,5)
Sensors: Multispectral Scanner (MSS): 1,2,3,4,5
Return Beam Vidicon (RBV): 1,2,3
Thematic Mapper (TM): 4,5

Landsats 1, 2, 3

MSS:

Band	Wavelength (μm)	Resolution (m)
4	0.5-0.6 (green)	80
5	0.6-0.7 (red)	80
6	0.7-0.8 (near IR)	80
7	0.8-1.1 (near IR)	80
8	10.4-12.6 (thermal) (3 only)	240

RBV:

Landsats 1,2

1	0.475-0.575 (blue)	80
2	0.580-0.680 (yellow)	80
3	0.690-0.830 (red)	80

Landsat 3

1	0.505-0.750	30

Landsat 4, 5

MSS:

1	0.5-0.6 (green)	80
2	0.6-0.7 (red)	80
3	0.7-0.8 (near IR)	80
4	0.8-1.1 (near IR)	80

TM:

1	0.45-0.52 (blue-green)	30
2	0.52-0.60 (green)	30
3	0.63-0.69 (red)	30
4	0.76-0.90 (near IR)	30
5	1.55-1.75 (near IR)	30
6	10.40-12.50 (thermal IR)	120
7	2.08-2.35 (near IR)	30

The Landsat satellites have been operating since 1972. A huge database of imagery has been collected with repetitive coverage that is useful for observing changes in the land over the 20-year span of the program. Landsats 4 and 5 are still functioning, but because they have exceeded their design lifetime they are not operated continuously. Landsat 6, which will carry an enhanced Thematic Mapper instrument, is scheduled for launch in early 1993. Landsat 7 is in planning stages.

Data and imagery available from:
U.S. Geological Survey
EROS Data Center, Customer Services
Sioux Falls, SD 57198 USA
(605) 594-6151
fax: (605) 594-6589

Earth Observation Satellite Corporation (EOSAT)
4300 Forbes Boulevard
Lanham, MD 20706 USA
(800) 344-9933

Large Format Camera, U.S.

Platform: Space Shuttle Mission 41-G
Operational Dates: October 5 to 13, 1984
Coverage: Selected sites between 57°N and 57°S
Altitude: 223-352 km
Sensor: ITEK Large Format Camera
Focal Length: 30.5 cm
Image Format: 22.9 x 45.7 cm
Resolution: 10 to 25 m, varying with altitude and film type

The Large Format Camera (LFC) flew aboard the Space Shuttle as part of an experiment to study terrain mapping with cameras from orbiting spacecraft. The 450-kg LFC produced high-resolution photographs covering large areas. It was operated from the Space Shuttle's cargo bay. A total of 2,143 photographs were taken with the LFC. Four film types were used: Kodak 3412 Panatomic-X Aerocon black and white, Kodak 3414 High Definition Aerial black and white, Kodak SO-131 High Definition Aerochrome infrared, and Kodak SO-242 Aerial color.

Photographic products available from:
U.S. Geological Survey
EROS Data Center, Customer Services
Sioux Falls, SD 57198 USA
(605) 594-6151
fax: (605) 594-6589

Metric Camera, Germany

Platform: Spacelab Mission 1 (STS-9)
Operational Dates: Nov. 28 to Dec. 8, 1983
Coverage: Selected sites between 57°N to 57°S
Altitude: 250 km
Camera: Metric Camera; Modified Zeiss RMKA 30/23 Aerial Camera
Focal length: 305 mm
Image size: 23 cm x 23 cm
Resolution: 20 m

The purpose of the Metric Camera experiment on Spacelab Mission 1 (flown on Space Shuttle Mission STS-9) was to see if imagery obtained from high-quality survey camera systems in low orbits could be used to make detailed maps of different regions of the world. Kodak 2443 Aerochrome color infrared film and Kodak 2405 Double-X Aerographic black-and-white film were used to produce 500 color infrared and 450 black-and-white photographs. The images were taken with at least 60 percent overlap for stereoscopic (three-dimensional) viewing.

Photographic products available from:
DLR-Forschungszentrum
Post Wessling/OBB
D-8031 Oberpfaffenhofen, Germany
(49)(081 53) 28-0
fax: (49)(081 53) 28-243

Marine Observation Satellite (MOS-1A & 1B), Japan

Operational Dates: MOS-1A: February 18, 1987 to present
MOS-1B: February 7, 1990 to present
Coverage: Coastal features between 81°N to 81°S
Altitude: 909 km
Sensors: Visible and Thermal Infrared Radiometer (VTIR)
Multispectral Electronic Self-Scanning Radiometer (MESSR)
Microwave Scanning Radiometer (MSR)

MESSR	Band	Wavelength (µm)	Resolution (m)
	1	.51-.59 (green)	50
	2	.61-.69 (red)	50
	3	.73-.80 (near IR)	50
	4	.80-1.10 (near IR)	50

VTIR	Band	Wavelength (µm)	Resolution (m)
	1	.5-.7 (visible)	.9
	2	6.0-7.0 (near IR)	2.7
	3	10.5-11.5 (thermal IR)	2.7
	4	11.5-12.5 (thermal IR)	2.7

MSR	Band	Frequency (GHz)	Resolution (km)
	1	31.4	23
	2	23.8	32

As their primary goals, the MOS satellites measure sea-surface temperature and color and serve as experimental platforms for Earth observation technology. The VTIR sensor scans the Earth's surface to detect suspended sediments in the visible range and water vapor in the near-infrared. In the thermal infrared wavelengths, the VTIR detects warm and cold currents in the oceans as well as urbanized features on land. Ocean color is monitored by the MESSR using two visible and two infrared sensors. The MSR monitors atmospheric water vapor, snow, and ice cover. Coverage of MOS-1 satellites is limited to regions with ground tracking and receiving stations.

Products available from:
NASDA/Earth Observation Center
1401 Numanove, Oohashi, Hatoyama-Machi
Hiki-Gun, Saitama 350-03 Japan
(81) 0492-96-1611
fax: (81) 0492-96-0217

NIMBUS-7, U.S.

Operational dates: Oct. 24, 1978 to present
Altitude: 950 to 1100 km
Coverage: Polar-orbiting on six-day repeat cycle for global coverage
Imaging sensors: Coastal Zone Color Scanner (CZCS)
Scanning Multichannel Microwave Radiometer (SMMR)
Total Ozone Mapping Spectrometer (TOMS)

	Band	Spectral Range (µm)	Resolution (km)
CZCS	1	.433-.453 (blue)	.825
	2	.510-.530 (green)	.825
	3	.540-.560 (yellow)	.825
	4	.660-.680 (red)	.825
	5	.700-.800 (near IR)	.825
	6	10.5-12.5 (infrared)	.825
TOMS	1	.312-.313	
	2	.3307-.3317	
	3	.317-.318	
	4	.3393-.3403	
	5	.3595-.3605	
	6	.3795-.3805	

		Frequency (GHz)	Resolution (km)
SMMR	1	6.6	97.5
	2	10.69	
	3	18.0	60
	4	21.0	
	5	37.0	30

Nimbus-7 is the latest of seven meteorological research satellites in the Nimbus series. All of the Nimbus satellites possessed sensors for obtaining atmospheric data. The Nimbus-7 introduced the Total Ozone Mapping Spectrometer (TOMS) and was also equipped with the CZCS and SMMR imaging sensors devoted to observing ocean dynamics.

The TOMS is the first sensor to provide global data coverage for the mapping of ozone distribution in the Earth's atmosphere. Collecting data in ultraviolet wavelengths, the sensor is able to measure ozone as well as various pollutants in the atmosphere. TOMS data are used to study ozone depletion in the atmosphere, most notably over Antarctica, and to measure sulfur dioxide released during volcanic eruptions.

The SMMR detects emitted microwave radiation under all cloud conditions. SMMR's data are used to measure sea-surface temperature, sea ice coverage, and near-surface wind speeds. The CZCS (shut down in 1986) measured radiation in the visible and infrared portions of the spectrum. CZCS data provide chlorophyll concentrations, sediment distributions, and sea-surface temperatures.

Photographic products for SMMR and TOMS imagery available from:
National Space Science Data Center
Code 933.4
NASA/Goddard Space Flight Center
Greenbelt, MD 20771 USA
(301) 286-6695
fax: (301) 286-4952

Photographic products for CZCS imagery available from:
NOAA/NESDIS/NCDC
Satellite Data Services Division
5627 Allentown Road
Camp Springs, MD 20746 USA
(301) 763-8400
fax: (301) 763-8443

Resource-01 and 02, Russia

Operational Dates: 1980 to present
Coverage: worldwide
Altitude: 650 km
Sensors: MSU-SK optomechanical scanner
MSU-E electronic optical scanner

	Wavelength Band (μm)	Resolution (m)
MSU-SK	0.5-0.6 (visible)	170
	0.6-0.7 (visible)	
	0.7-0.8 (near IR)	
	0.8-1.0 (near IR)	
	10.4-12.6 (thermal IR)	
MSU-E	0.5-0.6 (visible)	44
	0.6-0.7 (visible)	
	0.8-0.9 (near IR)	

The Resource-01 and 02 system includes three ground monitoring stations. Data can be collected in real time or from tape storage at stations in Moscow (western Russia), Novosibirsk (central Russia) or Khabarovsk (eastern Russia).

Images from these sensors are not currently commercially available. For information about the Resource satellites contact:
GLAVKOSMOS
9 Krasnoproletarskaya Street
103030 Moscow, Russia
(7) 258-22-20
telex: 411879 START

SEASAT, U.S.

Operational Dates: June 26, 1978 to Oct. 9, 1978
Coverage: Selected land and sea sites in North America, Central America, Western Europe, and the Caribbean
Altitude: 800 km
Imaging sensors: Synthetic Aperture Radar (SAR)
Visible and Infrared Radiometer (VIRR)

Band	Wavelength	Frequency	Resolution
SAR	23.5 cm	1.275 GHz	25 m
VIRR 1	.49-.94 μm		10.5 km
2	10.5-12.5 μm		

SEASAT was the first satellite designed specifically for observing the Earth's oceans. The SAR (Synthetic Aperture Radar) instrument provided radar images of the land and oceans and could do so regardless of weather and cloud cover conditions. Utilizing 5 microwave sensors, SEASAT obtained various forms of oceanographic data. In addition to the SAR and VIRR imaging sensors, SEASAT also carried a SEASAT-A Scatterometer System (SASS) to monitor ocean surface winds, a Radar Altimeter (ALT) to measure tides and wave heights, and a Scanning Multichannel Microwave Radiometer (SMMR) to observe sea-surface temperature, atmospheric water vapor, and sea ice.

Imagery available from:
NOAA/NESDIS/NCDC
Satellite Data Services Division
5627 Allentown Road
Camp Springs, MD 20746 USA
(301) 763-8400
fax: (301) 763-8443

Shuttle Imaging Radar (SIR-A & B)

Platforms: Space Shuttle Mission STS-2 (SIR-A)
Space Shuttle Mission 41-G (SIR-B)
Operational Dates: Nov. 12-14, 1981 (SIR-A)
Oct. 5-13, 1984 (SIR-B)

Coverage: Selected sites between 41°N and 35°S (SIR-A)
Selected sites between 57°N and 57°S (SIR-B)
Altitude: 259 km (SIR-A)
360 km (SIR-B)
Frequency: 1.278 GHz (SIR-A)
1.282 GHz (SIR-B)
Wavelength: 23.5 cm (SIR-A & B)
Resolution: 40 m (SIR-A)
20 m (SIR-B)

The Shuttle Radar instruments were carried and deployed in the Space Shuttle Cargo Bay. SIR-A imaged the Earth in strips 50-km wide and collected data on geologic and oceanographic features. Similarly, SIR-B operated over land and sea (with swath widths from 20- to 50-km), but added the capability to change the angle of the radar signal and to collect data in digital form.

Imagery available from:
National Space Science Data Center
Code 933.4
NASA Goddard Space Flight Center
Greenbelt, MD 20771 USA
(301) 286-6695
fax: (301) 286-4952

Space Shuttle Handheld Cameras, U.S.

Altitude: 204-555 km
Coverage: 28°N to 28°S or 57°N to 57°S
Cameras: Modified Hasselblad 70mm
Modified Linhof Aero Technika 45
35mm Nikon camera with charged-coupled device (CCD)

Astronauts on all Space Shuttle missions have carried along cameras, and most of the photographs they have taken are of the Earth. Earth observation is part of the training Space Shuttle crews receive before missions. During the flight, astronauts can choose places to photograph and mission planners alert the crew to good photo opportunities. Most Space Shuttle handheld photography has been in natural color (using various Kodak Ektachrome films), but there has also been color infrared photography (with Kodak Aerochrome film) and a small amount of black and white photography. The Linhof camera uses a larger film format than the Hasselblad and can show more detail. The Nikon camera with a CCD detector does not use film, but records images digitally. All flights have had a Hasselblad on board. Many selected flights also had a Linhof, while a CCD-equipped camera was first used on STS-44 in 1991.

Photographic products available from:
U.S. Geological Survey
EROS Data Center
Customer Services
Sioux Falls, SD 57198 USA
(605) 594-6151
fax: (605) 594-6589

SPOT 1,2, France

Operational Dates: SPOT-1—Feb. 1986 to Dec. 1990,
March 1992 to Oct. 1992
SPOT-2—Jan. 1990 to present
Coverage: 81°N to 81°S
Altitude: 830 km
Sensors: Two High Resolution Visible (HRV) Sensors—
panchromatic and multispectral modes

	Band	Wavelength (μm)	Resolution (m)
Panchromatic		.51-.73 (visible)	10
Multispectral	1	.50-.59 (green)	20
	2	.61-.68 (red)	20
	3	.79-.89 (near IR)	20

SPOT satellites have provided wide-ranging coverage of the Earth since 1986. The high-resolution sensors allow detailed monitoring of varied regions. SPOT-1 was called back to duty after more than a year's retirement to provide additional coverage over Europe. At the time of publication, both satellites are fully operational.

Digital tapes and photographic products available from:
SPOT Image Corporation
1892 Preston White Drive
Reston, VA 22091-4326
(703) 620-2200
fax: (703) 643-1813

SPOT Image
16 bis
Avenue Édouard Belin
B.P. 4359
31030 Toulouse Cedex, France
61-53-99-76

selected bibliography

Ahabriya, S. S. "Assessment and Mapping of Desertification in Rajasthan (India) Using Satellite Products." 20th International Symposium on Remote Sensing of the Environment, Dec. 4-10, 1986, 859-68.

Alvarez, W., and K. H. A. Gohrbandt, eds. Geology and History of Sicily. Petroleum Exploration Society of Libya, 1970.

Aveni, Anthony F. The Lines of Nazca. Philadelphia: American Philosophical Society, 1990.

Bair, Frank C., ed. Countries of the World and Their Leaders Yearbook. New York: Gale Research, Inc., 1991.

Baker, B. H. et al. "Geology of the Eastern Rift System of Africa." Geological Society of America, Special Paper 136 (1972).

Banks, Michael C. Greenland. Totowa, NJ: Rowman and Littlefield, 1975.

Barringer, F. "Chernobyl: Five Years Later the Danger Persists." New York Times Magazine, April 14, 1991.

Beaumont, P. "Salt Weathering on the Margin of the Great Kavir, Iran." Bulletin of the Geological Society of America 79 (1968): 1683-1684.

Beaumont, Peter, Gerald H. Blake, and J. Malcolm Wagstaff. The Middle East: A Geographical Study. New York: John Wiley, 1988.

Best, Alan C., and Harm J. DeBlij. African Survey. New York: John Wiley, 1977.

Blackford, Eric. "Space Shuttle Radar (SIR-A) Views Near East Volcanoes." Volcano News 17 (July 1984): 6-7.

Bromley, Rosemary D.F., and Ray Bromley. South American Development: A Geographical Introduction. Cambridge: Cambridge University Press, 1982.

Brookes, Ian A. "Above the Salt: Sediment Accretion and Irrigation Agriculture in an Egyptian Oasis." Journal of Arid Environments 17 (1989): 335-348.

Brooks, W. H., and K. S. D. Manor. "Vegetation Dynamics in the Asir Woodlands of Southwestern Saudi Arabia." Journal of Arid Environments 6 (1983): 357-362.

Browder, John O. Fragile Lands of Latin America. Boulder: Westview Press, 1989.

Cassidy, W. A. "Martian Gases in an Antarctic Meteorite?" Antarctic Journal of the United States 15,49 (1980): 651.

Cherfas, J. "East Germany Struggles to Clean Its Air and Water." Science 248 (1990): 295-296.

Chingchang, Biq. "Silk Road: A Natural Thoroughfare in the Continental Framework of Asia." Memoir of the Geological Society of China, 8 (April 1987): 293-300.

Clout, Hugh, Mark Blackstill, Russell C. Kine, and David Pinder. Western Europe: Geographical Perspectives New York: John Wiley, 1989.

Colwell, R. N., ed. Manual of Remote Sensing. Falls Church, Va: American Society of Photogrammetry, 1983.

Condie, Kent C. Plate Tectonics and Crustal Evolution. New York: Pergamon Press, 1982.

Cornillon, P. A Guide to Environmental Satellite Data. University of Rhode Island Technical Rept. 79. Rhode Island: University of Rhode Island, 1982.

Crow, John A. The Epic of Latin America. Berkeley: University of California Press, 1980.

Dann, Kevin T. Traces on the Appalachians. New Brunswick: Rutgers, 1988.

David, T. W. Edgeworth. The Geology of the Commonwealth of Australia: Volume II, London: Edward Arnold & Co., 1950.

Delpar, Helen. The Discoverers. New York: McGraw-Hill, 1980.

Douglas, Marjory Stoneman. The Everglades. Miami: Banyan Books, 1978.

Drewry, D.Y., ed. Antarctica: Glaciology and Geophysical Folio.Cambridge: Scott Polar Research Institute, 1983.

El-Baz, Farouk, ed. Deserts and Arid Lands. The Hague: Martinus Nishoff, 1984.

El-Baz, Farouk, and D.M. Warner, eds. Apollo-Soyuz Test Project Summary Science Rept. Washington, D. C.: NASA, 1979.

Europa Publications. The Middle East and North Africa. London: Europa Publications Ltd., 1987.

Fearnside, Philip M. "China's Three Gorges Dam: 'Fatal' Project or Step Toward Modernization?." World Development 16 (1988): 615-630.

Ferrigno, J.G., and W.G. Gould. "Substantial Changes in the Coastline of Antarctica Revealed by Satellite Imagery." Polar Record 23,146 (1987): 577-583.

Ford, J. P. et al. Seasat Views North America, the Caribbean, and Western Europe with Imaging Radar. JPL Publ. 80-67. Pasadena: NASA/Jet Propulsion Lab, 1980.

Ford, J. P., J. B. Cimino, and C. Elachi. Space Shuttle Columbia Views the World with Imaging Radar: The SIR-A Experiment. JPL Publ. 82-95. Pasadena: NASA/Jet Propulsion Lab, 1983.

Ford, J. P., J. B. Cimino, B. Holt, and M. R. Ruzek. Shuttle Imaging Radar Views the Earth: The SIR-B Experiment. JPL Publ. 86-10. Pasadena: NASA/Jet Propulsion Lab, 1986.

Frost, Peter. Exploring Cuzco. 3rd ed. Chalfont St. Peter: Bradt Enterprises, 1984.

Fu, Lee-Lueng, and Benjamin Holt. Seasat Views Oceans and Sea Ice with Synthetic Aperture Radar. JPL Publ. 81-120. Pasadena: NASA/Jet Propulsion Lab, 1982.

Gantz, C.O. A Naturalist in Southern Florida. Coral Gables, Florida: University of Miami Press, 1982.

Gellatly, Ann F., Trevor J.H. Chinn, and Friedrich Röthlisberger. "Holocene Glacier Variation in New Zealand: A Review." Quaternary Science Reviews 7 (1988): 227-242.

Grieve, R.A.F. et al. Astronaut's Guide to Terrestrial Impact Craters. LPI Technical Report 88-03. Houston: Lunar and Planetary Institute, 1988.

Griffiths, Ieuaun Ll. "The Scramble for Africa: Inherited Political Boundaries." The Geographical Journal 152, 2 (1986).

Grigor'ev, Al. A. "Large Scale Changes in Aral Coast Environment From Satellite Observations." Problemy Osvoeniya Pustyn 1 (1987): 16-22.

Grove, A.T. "The State of Africa in the 1980s." The Geographical Journal, 152, 2 (1986): 193-203.

Hack, J. T. "Dunes of the Western Navajo Country." Geographical Review (1941): 240-263.

Hammond. Past Worlds: The Times Atlas of Archaeology. Maplewood, NJ: Hammond, 1988.

Hecht, Susanna, and Alexander Cockburn. The Fate of the Forest. New York: Harper Collins, 1990.

Hemming, John. The Conquest of the Incas. New York: Harcourt Brace Jovanovich, 1970.

Hillary, Edmund. Nothing Venture, Nothing Win. Sevenoaks, Kent: Hodder and Stoughton, Ltd., 1975.

Hoffman, George W., ed. Europe in the 1990s: A Geographical Analysis. New York: John Wiley & Sons, 1989.

Hohenemser, C., M. Deicher, A. Ernst, G. Hofsass, G. Lindner, and E. Recknagel. "Chernobyl." Environment 28, 5 (1986): 6-42.

Holm, D. A. "Desert Geomorphology in the Arabian Peninsula." Science 132 (1960): 1369-1379.

Holmes, J.H., ed. "Queensland: A Geographical Interpretation." Queensland Geographical Journal 1, 4 (1986).

Holzer, Tom, and Audrey Salkeld. First on Everest. New York: Henry Holt and Company, 1986.

Hopley, David. "Anthropogenic Influences on Australia's Great Barrier Reef." Australian Geographer 19, 1 (1988).

Huntford, Roland. The Last Place on Earth. New York: Atheneum, 1985.

Innocenti, F., R. Mazzuoli, F. Pasquare, F. R. Brozolo, and L. Villari. "Evolution of the Volcanism in the Area of Interaction Between the Arabian, Anatolian, and Iranian Plates (Lake Van, Eastern Turkey)." Journal of Volcanology and Geothermal Research 1 (1976): 103-112.

Jennings, J.N., and J.A. Mabbutt. Landform Studies From Australia and New Guinea. Cambridge: Cambridge University Press, 1962.

Johns, R.K. "Investigation of Lake Gairdner." Report of Investigations no. 31, part 2. Geological Survey of South Australia.

Kane, Mouhamed Moustapha. "A History of Fuuta Tooro, 1890s-1920s, Senegal Under Colonial Rule, The Protectorate." PHD Dissertation. UMI, 1987.

Kelly, Donovan, and Don Finley. "Notes about the Armenia Earthquake of 7 December 1989." Earthquakes and Volcanoes 21, 2 (1989): 68-76.

Kennett, Audrey, and Victor Kennett. The Palaces of Leningrad. London: Thames and Hudson, Ltd., 1973.

Kennett, James. Marine Geology. Englewood Cliffs, N. J.: Prentice-Hall, 1982.

Kingma, Jacobus T. The Geological Structure of New Zealand. New York: John Wiley, 1974.

Kolata, Alan L., and Charles Ortloff. "Thermal Analysis of Tiwanaku Raised Field Systems in the Lake Titicaca Basin of Bolivia." Journal of Archaeological Science 16 (1989): 233-263.

Krinsley, D. B. "Geomorphology of Three Kavirs in Northern Iran." In Playa Surface Morphology: Miscellaneous Investigations, edited by J.T. Neal. USAF Office of Aerospace Research, Environmental Research Papers, No. 283, 1968.

Láng, I. "Hungary's Lake Balaton." Ambio 7, 4 (1978): 164-168.

Learmonth, Nancy, and Andrew Learmonth. Regional Landscapes of Australia. London: Heinnemann, 1971.

Leonard, H. Jeffrey. Natural Resources and Economic Development in Central America. New Brunswick: Transaction Books, 1987.

Levi, B.G. "Cause and Impact of Chernobyl Accident Still Hazy." Physics Today, July (1986): 17-21.

Lhote, H. "Oasis of Art in the Sahara." National Geographic 172 (1987): 181-191.

Lloyd, Peter C. The "Young Towns" of Lima. Cambridge: Cambridge University Press, 1980.

Löffler, E., and M. E. Sullivan. "Lake Dieri Resurrected: An Interpretation Using Satellite Imagery." Zeitschrift für Geomorphologie 23, 3 (1979): 233-242.

Lowman, Paul. "Geologic Orbital Photography: Experience from the Gemini Program." Photogrammetria 24 (1969): 77-106.

Mabbutt, J.A. "Landforms of the Australian Desert." In Deserts and Arid Lands, edited by F. El-Baz. The Hague: Martinus Nijhoff, 1984.

MacFadyen, J.T. "Houston's Seaway Shouldn't Work, Yet Somehow it Does." Smithsonian, Oct. 1985, 89-99.

Madigan C.T. "The Australia Sand Ridge Deserts." Geographical Review 26 (1936): 205-227.

Marples, D. R. "Ukraine Fallout Debate." Bulletin of Atomic Scientists, December (1989).

Marvin, Ursula B. "Meteorite Placer Deposits of Antarctica." Episodes 3, (1982): 10-15.

Maxwell, Ted, and Priscilla Strain. "Discrimination of Sand Transport Rates and Environmental Consequences in Central Egypt from SPOT Data." In SPOT-1 Image Utilization, Assessment, Results. (1987): 209-214.

McCarthy, T. S., I. G. Stanistreet, B. Cairncross, W. N. Ellery, K. Ellery, R. Oelofse, and T. S. A. Grobicki. "Incremental Aggradation on the Okavango Delta-Fan, Botswana." Geomorphology 1 (1988): 267-278.

McCauley, J. F., G. G. Schaber, C. S. Breed, and M.J. Grolier. "Subsurface Valleys and Geoarchaeology of the Eastern Sahara Revealed by Shuttle Radar." Science 218 (1982): 1004-1020.

McClelland, Lindsay, Tom Simkin, Marjorie Summers, Elizabeth Nielsen, and Thomas C. Stein, eds. Global Volcanism, 1975-1985. Englewood Cliffs, N. J.: Prentice-Hall, 1989.

McKee, E.D., ed. "A Study of Global Sand Seas." US Geological Survey Professional Paper 1052, 429. Washington D.C.: U.S. Government Printing Office, 1979.

McNaught, Kenneth A. The Pelican History of Canada. Harmondsworth, Middlesex: Penguin, 1982.

McSween, H. Y., Jr. "Clues to Martian Petrologic Evolution?" Reviews of Geophysics 23,4 (1985): 391-416.

Milliman, J.D., and Jin Qingming, eds. Continental Shelf Research—Sediment Dynamics of the Chang Jiang Estuary and the Adjacent East China Sea 4, 1/2 (1985).

Miyakawa, Yasuo. "Metamorphosis of the Capital and Evolution of the Urban System in Japan." Ekistics 50 (1983): 110-122.

Morris, Arthur S. South America. 3rd ed. London: Hodder and Stoughton, 1987.

Nagashima, Catherine. "The Tokaido Megalopolis." Ekistics 48 (1981): 280-301.

Neal, James T., ed. Playas and Dried Lakes. Stroudsburg, PA: Dowden, Hutchinson & Ross, 1975.

Newell, N. D. "Geology of the Lake Titicaca Region, Peru and Bolivia." Geological Soc. Amer. Mem. 36 (1949).

Newhall, B. The Airborne Camera. New York: Hastings House, 1969.

Nuclear Energy Agency-OECD. The Radiological Impact of the Chernobyl Accident. Paris: OECD, 1987.

Nyrop, Richard F., ed. "Saudi Arabia, A Country Study." Foreign Area Studies (1984). Washington: The American University.

Ollier, C. Volcanoes. Oxford: Basil Blackwell Ltd., 1988.

Painton, F. "Where the Sky Stays Dark." Time, 28 May 1990, 40-42.

Pfeffer, Pierre. Asia, A Natural History. New York: Random House, 1968.

Pirkle, E. C., and W. H. Yoho. Natural Regions of the United States.Dubuque: Kendau/Hunt, 1982.

Polmar, Norman. Guide to the Soviet Navy. Annapolis: U. S. Naval Institute, 1983.

Polo, Marco. The Travels. Translated by Ronald Lathan. New York: Penguin, 1958.

Powers, R. W., L. F. Ramirez, C. D. Redmond, and E. L. Elberg, Jr. "Geology of the Arabian Peninsula: Sedimentary Geology of Saudi Arabia." USGS Professional Paper 560-D, 147. Washington DC: U.S. Government Printing Office, 1966.

Price, R. J. Glacial and Fluvioglacial Landforms. New York: Hafner Publishing Co., 1973.

Sader, Steven A., George V.N. Powell, and John H. Rappole. "Migratory Bird Habitat Monitoring Through Remote Sensing." International Journal of Remote Sensing 12, 3 (1991): 363-372.

Sborshchikov, I.M., L.A. Savostina, and L.P. Zonenshain. "Present Plate Tectonics Between Turkey and Tibet." Tectonophysics 79 (1981): 45-73.

Schell, Orville. "Journey to the Tibetan Plateau." Natural History 91 (Sept. 1982): 48.

Schmidt, W.A. "Are Great Lake Fish Safe to Eat?" National Wildlife, Ag/Sp 1989, 17-19.

Schreider, Helen, and Frank Schreider. Exploring the Amazon. Washington, D.C.: National Geographic Society, 1970.

Seely, Mary K. "The Namib Dune Desert: An Unusual Ecosystem." Journal of Arid Environments 1 (1978): 117-128.

Sharp R.P. Living Ice. Cambridge: Cambridge University Press, 1989.

Shimer, John A. Field Guide to Landforms in the United States. New York: Macmillan, 1972.

Short Nicholas M., and L.M. Stuart. The HCMM Anthology. NASA SP-465. Washington D.C.: NASA, 1982.

Short, Nicholas M., and Robert W. Blair, eds. Geomorphology From Space. Washington, D. C.: NASA, 1986.

Simkin, T. et al. Volcanoes of the World. Stroudsburg, Pa.: Hutchinson Ross Publishing Co., 1981.

Sit, Victor F., ed. Chinese Cities—The Growth of the Metropolis Since 1949. Hong Kong: Oxford University Press, 1985.

Sivin, Nathan, ed. The Contemporary Atlas of China. Boston: Houghton Mifflin, 1988.

Smil, Vaclav. The Bad Earth: Environmental Degradation in China. Armonk, N. Y.: M. E. Sharpe, 1984.

Soons, J. M., and Selbly. Landforms of New Zealand. Auckland: Paul Longman, 1982.

Strain, P. L., and F. El-Baz. "Sand Distribution in the Kharga Depression of Egypt." In Proc. Intn'l Symp. on Remote Sensing of Environment—1st Thematic Conference, (1982): 765-74.

Swithinbank, C. "Ice Movement of the Valley Glaciers Flowing into the Ross Ice Shelf, Antarctica." Science 141 (1963): 523-524.

Swithinbank, C. "Satellite View of McMurdo Sound, Antarctica." Polar Record 16, 105 (1973): 851-854.

Swithinbank, C. "To the Valley Glaciers that Feed the Ross Ice Shelf." The Geographical Journal 130, 1 (1964): 32-48.

Symons, Leslie, ed. The Soviet Union, A Systematic Geography. Barnes & Noble, 1990.

Thesiger, Wilfred. Arabian Sands. London: Longman Green, 1959.

Thesiger, Wilfred. The Marsh Arabs. New York. Dutton, 1964.

Thompson, Jon. "East Europe's Dark Dawn." National Geographic 179, 6 (1991): 37-63.

Turnock, D. The Human Geography of Eastern Europe. London: Routledge, 1989.

U.S. General Accounting Office. Growth and Use of Washington Area Airports. Washington D.C.: GAO, 1971.

Walker, A.S. "Deserts of China." American Scientist 70, (1982): 366-376.

Wallach, P. "Dark Days." Scientific American, August (1990): 16-17.

West, Robert C. et al. Middle America: Its Lands and Its Peoples. Englewood Cliffs, NJ: Prentice-Hall, 1989.

Weyl, Richard. Geology of Central America. 2nd ed. Berlin: Gebrüder Borntraeger, 1980.

Williams, Lynden Starr. The Suburban Barriadas of Lima: Squatter Settlements as a Type of Peripheral Urban Growth in Peru. Ph.D. Diss. California State College, 1969.

Williams, R.S. et al. "Documentation on Satellite Imagery of Large Cyclical or Secular Changes in Antarctic Ice Sheet Margin." EOS 69, no. 16 (1988): 365.

Williams, R. S, and J. G. Ferrigno, eds. Satellite Image Atlas of Glaciers of the World. USGS Professional Paper 1386. Washington, D.C.: U. S. Government Printing Office, 1988.

Wilson, R. "A Visit to Chernobyl." Science 236 (1987): 1636-1640.

Young, Margaret Walsh, and Susan L. Stetler, eds. Cities of the World. Detroit: Gale Research Co., 1983.

Zeil, Werner. The Andes: A Geological Review. Berlin: Gebrüder Borntraeger, 1979.

Zierer C.M. "Brisbane—River Metropolis of Queensland." Economic Geography 17, 4 (1941): 325-344.

Zwally, H. J., J. Campbell, F. Carsey, and P. Gloersen. Antarctic Sea Ice 1973-1976: Satellite Passive Microwave Observations, SP-459. Washington, D.C.: NASA, 1983.

acknowledgments

This work grew from an exhibition in the National Air and Space Museum called "Looking at Earth," which opened in 1986. The exhibition was sponsored in part by a contribution from Eastman Kodak Company. Sponsorship has continued through a generous grant from Eastman Kodak Company, Aerial Systems, for the research and image acquisition needed to publish this book. The National Air and Space Museum gratefully acknowledges Kodak's continued support of our efforts in science and geographic education. Special thanks are due to Len LaFeir and Dave Woods for their hard work and support. Work on this book was also supported through the Smithsonian Institution's Office of Quincentenary Programs, ably and graciously directed by Alicia Gonzales.

Many organizations contributed imagery to Looking at Earth. We are very grateful for their generosity and for the assistance of the following individuals: Barbara Backhaus and Manfred Schroeder of DLR-Forschungszentrum in Germany, Jim Balcerski and Lou Belcher of the Environmental Research Institute of Michigan (ERIM), Luis Bartolucci of the Mid-America Remote Sensing Center (MARC), George Büttner of Földmérési és Távérzékelési Intézet in Hungary, Robert M. Carey of the National Oceanographic and Atmospheric Administration's National Environmental Satellite Data and Information Service and National Climate Data Center (NOAA/NESDIS/NCDC), Judy Carroll of IMAX Systems Corporation in Canada, Kevin Corbley of the Earth Observation Satellite Corporation (EOSAT), Roberto Pereira da Cunha of Instituto Nacional de Pesquisas Espaciais (INPE) in Brazil, Jane Ferrigno of the U.S. Geological Survey, Peter Gottfried of Natural Systems Analysts, Inc., Hiroshi Kikuchi of the National Space Development Agency of Japan (NASDA), O. V. Kuleshov and Constantin Ovchinnikov of GLAVKOSMOS in Russia, Lynne Potter Lord and Wade Stewart of MacDonald Dettwiler in Canada, Baerbel K. Lucchitta of the U.S. Geological Survey, Yashushi Muranaka of the Remote Sensing Technology Center of Japan (RESTEC), Funso Olujohungbe of the U.N. Economic Commission for Africa, Wayne Rohde of the Earth Resources Observation Systems (EROS) Data Center, Jim Riordan of the National Space Science Data Center (NSSDC), Steven Sader of the University of Maine, Greg Scharfen of the National Snow and Ice Data Center, Jon Schneeberger of the National Geographic Society, Thomas Sever of the NASA/Stennis Space Center, Yoshio Tanaka of Tokai University in Japan, and Richard S. Williams, Jr., of the U.S. Geological Survey. Special thanks to Clark Nelson of SPOT Image, whose extra efforts on our behalf are greatly appreciated. Space Shuttle photography was supplied by the NASA/Johnson Space Center and the Lunar and Planetary Institute. Joanne Heckman provided extensive and knowledgeable research on image acquisition across the globe.

Staff and interns at the National Air and Space Museum's Center for Earth and Planetary Studies (CEPS) made important contributions to this work. The support and guidance of the chairman, Thomas R. Watters, are always appreciated. Andrew Johnston, with skill, hard work, and ingenuity, prepared the base maps and the map computer database. Mapping and appendix materials were also provided through the dedicated and talented work of Victoria Portway, Jennifer Pulos, and Gregory Coleman. Donna Slattery supplied secretarial support. Her patient and good-natured efficiency is an asset to any project. Rose Steinat (photo librarian), Michael J. Tuttle (geologist), and Anthony Ayers (volunteer) also provided invaluable assistance. Former staffers Constance Andre and Donald Hooper contributed materials for the introduction. Special mention goes to Ted A. Maxwell, the National Air and Space Museum's senior advisor for science, who first conceived and set the groundwork for this book.

Thanks are also due to Smithsonian staff in other departments, including Patricia J. Graboske, chief of publications, and Susan Beaudette, development officer, at the National Air and Space Museum. Innumerable literature searches and interlibrary loans were conducted on our behalf by Amy Levin, Mary Pavlovich, and Dave Spencer of the NASM Branch of the Smithsonian Institution Libraries. The advice and assistance of Daniel Cole, Smithsonian GIS coordinator, are greatly appreciated. Lindsay McClelland and David Lescinsky of the National Museum of Natural History's Global Volcanism Program helpfully answered our numerous volcano inquiries.

The authors greatly appreciate the assistance of Aldo Barsotti, Gordon Austin, Ed Chin, George Coakley, David Doan, and Pui-Kwan Tse of the U.S. Bureau of Mines. We are very grateful for the assistance of George Powell of the National Audubon Society. Our thanks also to Ian McLeod and Colin J. Simpson of the Australian Bureau of Mineral Resources, Geology, and Geophysics and to C. B. Smith of CRA Exploration Pty., Ltd., for all their assistance with our questions about Australia.

We are grateful for the work of the Review Panel: Luis Bartolucci of MARC (Middle America), David Doan of the U.S. Bureau of Mines (Pacific Islands), Farouk El-Baz of Boston University's Center for Remote Sensing (Middle East), Patricia Jacobberger of CEPS (Africa), Ted A. Maxwell of CEPS (Africa), Julian Minghi of the University of South Carolina (Europe), Colin J. Simpson of the Australian Bureau of Mineral Resources, Geology, and Geophysics (Australia), Christopher Smith of SUNY Albany (Asia), Richard S. Williams, Jr. of the USGS/Woods Hole (Antarctica), Charles A. Wood of the University of North Dakota (Africa, Middle America), and Jim Zimbelman of CEPS (North America). Their thoughtful reviews and insights were a valuable contribution to the work. Initials at the end of individual image descriptions indicate that the caption was edited from text contributed by those listed below. We are very grateful to these colleagues for lending their expertise to this project.

Turner Publishing acknowledges the special help of the following people: James Porges, Laura Heald, Terry Davila, Vivian Lawand, Rhonda Myers, Larry Larson, Virginia Pirie, Peri Koch, Lori Jones, Jane Lahr, Robin Aigner, Marcy Baron; the staff at Graphics International, including David Allen, J.C. Poole, Wes Aven, Lisa Davis, Jim Kennedy, Claudia McCue, Vicki Rumley, Tina Turner, and Kim Kiser; the staff at R.R. Donnelley & Sons, including Steve Neely, Debby Turoff, Bob Gospodarek, and Jerry Fletcher; and the staff at McQuiddy Printing, including Steve Sermonet, David McQuiddy Jr., and Hal Rehorn.

The following people, addressed by initials in the book, contributed text:

R. A. C.	Robert A. Craddock Center for Earth and Planetary Studies National Air and Space Museum
D. B. D.	David Bentley Doan Division of International Minerals Bureau of Mines, U.S. Department of Interior
R. D. F.	Randall D. Forsythe Department of Geography and Earth Science University of North Carolina, Charlotte
D. M. H.	Donald M. Hooper Department of Geology SUNY at Buffalo
P. A. J.	Patricia A. Jacobberger Center for Earth and Planetary Studies National Air and Space Museum
T. B. J.	Thomas B. Jones Earth Satellite Corporation
J. F. L.	Jim F. Luhr Department of Mineral Sciences National Museum of Natural History
T. A. M.	Ted A. Maxwell Center for Earth and Planetary Studies National Air and Space Museum
T. R. W.	Thomas R. Watters Center for Earth and Planetary Studies National Air and Space Museum
C. A. W.	Charles A. Wood Space Studies University of North Dakota, Grand Forks
J. R. Z.	James R. Zimbelman Center for Earth and Planetary Studies National Air and Space Museum